T0383785

Cancer and AIDS

Christopher Kwesi O. Williams

Cancer and AIDS

Part IV: Future Perspectives

Christopher Kwesi O. Williams
Hematology Oncology Consultancy
Port Angeles, WA, USA

ISBN 978-3-319-99237-2 ISBN 978-3-319-99238-9 (eBook)
https://doi.org/10.1007/978-3-319-99238-9

Library of Congress Control Number: 2018959099

This Springer imprint is published by the registered company Springer Nature Switzerland AG
The registered company address is: Gewerbestrasse 11, 6330 Cham, Switzerland

Foreword

The reader will find in the pages of this book an extensive and thought-provoking review by an experienced oncologist, with a long and deep experience in Africa, Europe, and North America, and an astute observer of the international scene. The focus is on two very different disease complexes: HIV/AIDS and cancer. However, although they have different pathogenesis, epidemiology, and therapy, they share many similarities in the problems, which they pose to health systems in low- and middle-income countries (LMIC). The author reminds of the difficulty of providing affordable health care in those parts of the world subsumed by the label of "LMICs" (and indeed of the fact that, within this group, incomes and resources also vary enormously). The fact that, within the health sector alone, the governments of these countries are facing the double burden of an increasing load of noncommunicable disease, while the traditional problems on infections/maternal and child mortality, remain. Basically, the issue is how low-income countries can address these challenges with the resources available. Of course, addressing the root cause (of inequalities of opportunity and wealth around the world) might be the logical approach; in this book, we cannot expect solutions to redressing the world economic order (where primary producers are rewarded less than manufacturers and they in turn less than "service providers"), rather, how to make do within this framework. Low income (exacerbated by gross inequalities in its distribution) results in poor health-care provision from public sources (government or social security schemes), with correspondingly poor results. Time and again, the author makes reference to the standards and guidelines developed in high-income countries (especially the USA) and asks how can they be made relevant to low-income settings. Of course, almost always they cannot.

What to Do?

Currently, the focus is upon action plans for NCDs, as sparked by the Declaration the United Nations General Assembly on the Prevention and Control of Noncommunicable Diseases [1], which the WHO followed up with its global monitoring framework on NCDs [2]. The latter urges the establishment of multisectoral national NCO plans. Is this a sensible idea from the point of view of implementation? In fact, although NCDs have some commonalities (for instance, they are not communicable), from a diagnosis and management point of view, there is nothing in common between, for example, diabetes, hypertension, and cancer. It is true that in high-income settings, some NCDs share common risk factors (tobacco for some cancers and heart disease and obesity for diabetes and some cancers), but in low-income settings, there is in reality little overlap. Indeed, as extensively documented in Chaps. 3 and 6 of Part II, the most important cause of cancer in LMICs is infection (especially with HPV, hepatitis viruses, *Helicobacter pylori*, and HIV itself). The reality is that the control strategies for cancers (embracing surveillance, prevention, early detection treatment, and palliation) are generally quite different from those for other NCDs.

How to develop some sort of plan to "control" cancer, given limited infrastructure and resources? There are many clues and suggestions throughout the book. The author writes in Chap. 9 of Part III: "The complexity of modern cancer management could be so overwhelming, especially for cancer caregivers of low-income countries, that the practice of cancer control tends to promote its prevention in preference to its management. However, a reasonable balance between the various aspects of cancer control is probably more reasonable." This is surely correct. A previous director of the Cancer Unit of WHO used to castigate hospitals providing cancer care as "white elephants," a charge that was grossly unfair to their staff. Care of the sick is an unavoidable minimum for any health-care system. Prevention may well be more logical, and cheaper, but this is of little comfort to those struck down with disease. And, the author reminds us in Chap. 7 of Part II, of the most important factor in determining who will get cancer-chance. Rediscovered, recently [3], in antiquity, Job discovered that a blameless past will not protect one from disaster (brought on by the unknowable will of a divinity or by a sequence of random mutations), although it is of course possible to change one's odds of disease through appropriate preventive action.

Cancer control, then, involves a balance of prevention, early detection and care (curative and palliative), with the balance determined by needs, resources, and the efficacy of different interventions. And, intervention is needed – some of the epidemiological transitions described in the book (such as the decline in incidence of cancers of the cervix and lung) are not natural phenomena, like the seasons, but the result of active interventions. Looking at priorities for cancer prevention, it is tempting to look to the guidelines of prestigious bodies, especially in the USA (see Chap. 8 of Part III). These will almost always be quite inappropriate to the task in hand. Each unit (country) will need to examine its own cancer profile, and the prevalence of risk factors, to quantify the fractions of cancer that is preventable, before weighing up the feasibility of doing so. The author points out the bizarre fact that many countries are contemplating the more costly (and difficult) proposition of vaccination against HPV (and being urged to vaccinate boys as well as girls), while incorporation of the HBV vaccine into infant immunization schedules is incomplete.

The head of the UK Screening Evaluation Unit began a lecture by noting that, among the strategies for controlling cancer, screening was the least important. She was surely correct. The logic of screening and the technological wizardry involved make it almost irresistible to cancer experts. Yet, as the author points out, in Chap. 8 of Part III, "for screening to be effective, it has to be population-based, whereby each person in the eligible population is invited to attend each round of screening.... this involves the establishment of a national public policy documented in a law, or an official regulation decision directive or recommendation. Although standard in resource-rich countries, this is not practicable in countries of limited resources, for economic reasons." As a result, cancer screening in low-income settings has generally been limited to local, opportunistic projects (e.g., detection of preinvasive cervical lesions, using VIA), with unknown, but surely very limited, impact on the population. There are much more compelling reasons to work out how to improve stage at presentation of cancer, which, as the discussion in Chap. 8 of Part III documents, means acting at the individual, community, and system (health service) levels.

Dealing with the most appropriate treatment services to offer is perhaps the most difficult part of cancer control, but it cannot be avoided, and the issues are fully discussed in Chap. 9 of Part III. Radiotherapy, the most expensive of interventions (in terms of equipment, trained personnel, maintenance, and management), is paradoxically the most required in low-income settings where patients present with advanced disease and palliative care is essential. The cost of drugs, especially the newer targeted immunotherapeutic drugs, is a huge concern. There is no alternative but to fall back on some sort of cost-per-life-year approach, which is implicitly the basis of the essential medicines program and more explicitly of some national regulatory agencies.

Palliative services get appropriate recognition as an essential component of cancer control. Really, no one should die in pain when the remedy is so cheap, making it always available must be a top priority for any "care" service. What of research and science, a field the author knows well? As he points out, it is another area of inequality; not only have the brain drain but also the lack of resources hampered research in the health sciences in LMICs. There is a lack of even the most fundamental research into health service need and performance. Look at the evidence the author could assemble of basic measures of cancer prognosis and outcome – for the USA, the SEER survey allows precise information on cancer stage and prognosis, their distribution and trends; for lower-income countries, the author must fall back on miscellaneous clinical series from local journals, with who-knows-what relevance to the population scenario. Too many of the articles cited seem to be commentary or diagnosis ("look at the problems" with far too little practical basis and providing no investment in legacy for the future). The Bill & Melinda Gates Foundation (*impatient optimists working to reduce inequity*) has given $279 million to the University of Washington study disease patterns worldwide. How much of this will be spent in improving the means to collect and analyze such important data in LMICs, where they are sorely lacking?

It is appropriate, given this excess of hand-wringing articles that the author in his final chapter provides some ideas for moving forward. The situation is unlikely to improve spontaneously; for every Singapore, there is a Somalia. His prescription of international action is surely the best hope of allowing the transfer of wealth and expertise from rich to poor. Just as climate change has forced reluctant international coordinated action, we must aim for the same in health. Sporadic efforts by a myriad of self-appointed NGOs are likely to be as effective as their efforts in conflict zones, where lack of coordination leads to as much effort going into interagency completion as to the task in hand. The model of focused partnerships in the Global Fund to Fight AIDS, Tuberculosis, and Malaria (GF) seems highly relevant. Global Health System strengthing? One can only hope that the UN and its specialist agencies can seize the opportunity; action is surely overdue!

United Nations General Assembly: Political declaration of the high-level meeting of the general assembly on the prevention and control of non-communicable diseases, UN New York, 2011 [available at http://www.un.org/en/ga/ncdmeeting2011/].

Honorary Senior Research Fellow
CTSU, Richard Doll Building
Old Road Campus
Roosevelt Drive
Oxford OX3 7LK, UK

Donald Maxwell Parkin

References

1. World Health Organization. Global action plan for the prevention and control of noncommunicable diseases 2013–2020. 2013. Available at: http://apps.who.int/iris/bitstream/10665/94384/1/9789241506236_eng.pdf
2. Tomasetti C, Vogelstein B. Cancer etiology. The number of stem cell divisions can explain variation in cancer risk among tissues. Science. 2015;347:78–81.
3. http://www.washington.edu/news/2017/01/25/bill-melinda-gates-foundation-boosts-vital-work-of the-uws-institute-for-health-metrics-and-evaluation/

Preface

The world can be subdivided into different categories, depending on the nature of the characterization. Perhaps, the best instrument of characterization of the populations of the world is the United Nations Human Development index (HDI), which classifies countries into "very high," "high," "medium," and "low" ranks of development, based on a variety of criteria. The topmost 49 of about 170 countries with the highest HDI scores are classified as being "very highly" developed. Many countries in this category are of almost unlimited human and financial resources. They contain less than 20% of the world population. Not only do the less-developed parts of the world harbor the greater burden of cancer, because, partly, that is where the majority of the world population lives. The predominance of the world's retroviral infectious burden, including HIV/AIDS and HTLV-1/HTLV-2, in these areas further compounds the nature and challenges of health care there.

Much of the international literature on cancer covers the nature and challenges of the disease and its control from the point of view of the high-income regions of the world. This is because of the presence in this region of mature and well-structured health-care systems. Doing so, however, gives a skewed view of cancer for the whole world. As the low- and middle-income regions of the world transition from communicable to noncommunicable disease patterns, however, there is a need for a corresponding paradigm shift. Cancer control measures of the high-income countries are largely impracticable in low-income countries, because they are simply not affordable there. The questions then arise as to whether cancer control should be a prerogative of high-income regions of the world or how this can be accomplished in the low- and middle-income settings as well. These are some of the questions that need to be addressed if a reduction of the sufferings caused by premature death from cancer and HIV/AIDS in the prime of life in much of the world is to be curtailed. This is the goal that this book aims to achieve.

The book provides a description of the epidemiology of cancer and retroviral diseases, including HIV/AIDS, with special reference to resource-poor settings, based on the author's own observations. For example, even though the author's background is adult medical oncology, he was preoccupied while working in the 1980s in Nigeria with childhood malignancies, especially Burkitt lymphoma, because adult malignancies were much less common in the hospital settings. Childhood acute lymphoblastic leukemia, which he encountered very frequently during his medical training and practice in high-income countries, was much less frequent in Nigeria, while cases of childhood acute myelogenous leukemia were, to his astonishment, commonly associated with mass formation (chloromas) at presentation. These observations were among the reasons for the author's interest in the role of lifestyle and environmental factors in pathogenesis of various childhood and adult cancers as outlined in the book. Furthermore, the difficulties that he encountered in providing appropriate care to a vulnerable segment of the community provoked in him a passion to find ways to address health-care system deficiencies in cancer control.

The author of this book is uniquely positioned to address the global challenges in the control of cancer and retroviral diseases, because of his global education and academic medical practice. Born and raised in Nigeria, he had his medical education in Munich, Germany. He subsequently underwent postgraduate education in Canada and the United States, followed by academic medical practice in Africa, Europe, and the Middle East, including extensive research-related travels in India, Brazil, and Argentina. Decades of practice of hematology and medical oncology in the United States and Canada, including years of service as a principal investigator of the National Cancer Institute of Canada Clinical Trials Group (NCIC-CTG) have given him a rich experience of the world's leading health-care systems. The beginning of his training in medical oncology in New York City under the tutelage of some of the authors of the blueprint of the United States "National Cancer Act," which President Nixon signed into law in 1971, a few years earlier, places him at a vantage point of observing the evolution of cancer control in the many decades of unprecedented advances that have followed. Similarly, his early involvement in human retroviral research in Africa, through his collaboration with leading scientists of the National Cancer Institute, in Bethesda, MD, USA, beginning at a point in time prior to the recognition of the human immunodeficiency virus (HIV) as the causative agent of the acquired immunodeficiency disease (AIDS), has also given him the opportunity to follow the evolving human tragedy of HIV/AIDS pandemic and its impending end. International exposures through his over 35-year membership of prestigious cancer control organizations, including the American Society of Clinical Oncology (ASCO) and the American Association for Cancer Research (AACR) have given him the opportunity to follow the evolution of the science of oncology and virology. His commitment to the elucidation of cancer control challenges in Africa and other developing parts of the world is what drove him to join others as a co-founder in 1982 of the African Organisation for Research and Training in Cancer (AORTIC).

In an era, in which there is a genuine concern for global equity in access to health, this book hopes to serve those who seek to understand the forces that shape global health-care systems, what needs to be done in the LMICs, where help is needed. These include health-care practitioners of all health-care systems, especially those in the "very developed" countries who are interested in global health care as a career. It will prove useful for funding agencies in the "very developed" countries providing assistance to health-care providers, researchers and others in the less developed world. It will hopefully also be a valuable resource for health-care providers and health care policy-makers in resource-poor settings of the world, who seek to understand the dynamics of health-care provision in all health-care systems.

Port Angeles, WA, USA Christopher Kwesi O. Williams

Contents

Part IV
Future Perspectives

Chapter 10
Strategies for Progress

Abstract The future of cancer and HIV/AIDS control in low- and middle-income countries (LMICs) lies in strategic international collaboration for opportunities in global equity in health, so as to mitigate high rates of premature death and suffering. This goal should be attainable through successful performance of the United Nations initiatives, including the Sustainable Development Goals, and performing in concert with funding agencies, such as the Global Fund, and health care promoting agencies including the Global Alliance for Vaccine Development, the United States National Institutes of Health and its subsidiaries, such as the Fogarty International Center. Other organizations already making remarkable contributions in global health (GH) outreach are Médicins Sans Frontières, the Bill and Melinda Gates Foundation and several other mainly US-based research agencies, working "in convergence" with academia and industry to resolve GH care challenges. Promotion of emerging GH partnerships between cancer care organizations in high-income countries (HICs) and LMICs promises to impact positively on the control of cancer and HIV/AIDS globally. Innovations are required in promoting resource-setting appropriate access to universal health coverage and supportive human resources development, while exploring ways to engage the political class to comprehend the basis of, and need for the control of factors that are responsible for premature death and suffering among their people. LMICs need vigorous advocacy of science as an instrument of human development, for the conquest of poverty and the containment of the ravages of diseases, so as to ensure the global dissemination of the US cancer Moonshot, and UNAID 90-90-90 goals.

Keywords Global fund · Moonshot · UNAIDS · UHC · 90-90-90 · Convergence · Global health · GAVI · NCCN · Obio

C. K. O. Williams, *Cancer and AIDS*,
https://doi.org/10.1007/978-3-319-99238-9_1

10.1 The Role of International Partnership in the Development of Global Health Systems

Today, health care systems are dramatically varied around the world, as discussed and illustrated in Chap. 9 in Part III (see Table 9.1 in Part III). Conversely, health care systems in all countries, rich and poor, play a bigger role in people's lives than ever before [1]. These are relatively recent developments. For thousands of years, traditional practices, often integrated with spiritual counseling were the prevalent methods of providing both preventive and curative health care in all cultures, and continue to coexist even today with modern medical practices. About 100 years ago, organized health systems in the modern sense barely existed [1]. Most recently, there has emerged the realization that appropriate investment in health is associated with enormous dividend, which could translate to dramatic health gains by 2035. The instrument for this achievement is what has been termed "progressive universalism", a pathway to universal health coverage (UHC), an efficient way to achieve health and financial protection [2].

The UN High-Level Meeting (UN HLM) of September 2011, which was convened to address the challenges of non-communicable diseases that were recognized as threatening economic and human developments led to initiatives, which would not only contribute to their control, but also benefit selected Millennium Development Goals (MDGs) [3]. The public-private partnership inspired Global Fund to Fight AIDS, Tuberculosis, and Malaria (GF), launched in 2001, has not only played a major role in promoting advances in evidence-based public health efforts in developing countries between 2001 and 2016, it has the potential of serving as a template for further funding of Sustainable Development Goals (SDGs), the UN initiative to succeed MDGs after its expiration in 2016. The Global Fund, an initiative of the then UN Secretary-General, Kofi Annan, was endorsed by the leaders of the Group of Seven (G7) countries, and led to the disbursement of $35 billion, accounting for 16.4% of international funding for HIV/AIDS, 44.5% for TB, and 81.4% for malaria bed nets [4, 5]. "By the end of 2014, GF-supported programs had provided 8.1 million people with antiretroviral drugs (ARVs), distributed 548 million bed nets, treated 515 million people with artemisinin-based combination therapy, and treated 13.1 million people for TB [5, 6]. Eight design principles (DP) have made the Global Fund an innovative financial institution. They include: [7] country-led; [2] multistakeholder; [3] independent, transparent, technical review, and evaluation; [4] political independence; [5] needs-based pooled financing; [7] funding is for disease-specific programs but is implemented in broader health systems; [7] performance based funding; and [8] financing only. The #8 design principle ensures that GF, which is essentially a donor institution, like the World Bank, functions as a financing-only instrument rather than an implementing agency, and relies on technical partners, e.g. the World Health Organization and the UN Programme on HIV/AIDS, for technical advice or implementation support to countries [5].

GF has a track record through its MDG-era performance in the in promoting worthy projects to actualization, a process that had been deemed impossible in resource-poor settings, due to perceived inefficiency to execute complex health interventions emanating from poor countries. The #3 design principle has enabled the identification of high-quality, country-led proposals designed to operate in complex operating environments of fragile countries [8]. DP #1, #3, #4 and #5 had taught countries that rigorously designed projects would receive independent, transparent technical reviews and evaluation leading to successful funding by GF [5, 9]. The GF-funded implementation research has been adopted by WHO as standard for some commodities (DP #8) [10], and can be used in the free mass distribution of essential medicines, such as antiretroviral and anticancer agents. Thus, GF funding mechanism could serve as in cancer and HIV/AIDS control strategies, thus, enabling the UNAIDS project 90-90-90 for the eradication of HIV/AIDS by 2030 (see Sect. 9.14 in Part III), curative management of cancer in specific populations using the WHO essential medicines, or in palliative care in low- and middle-income countries (see Chap. 9 in Part III). Furthermore, the GF-based funding and the associated design principles can help in accelerating innovation and propagation of best practices in health care in low- and middle-income countries [2], while contributing to health services strengthening where it is required [11, 12]. By so doing, GF-based funding can contribute immensely to obliterating the disparity that exists in the health care standards between the high-income countries and low- and Middle-income countries, in general, and in particular, in respect of the burdens of cancer and HIV/AIDS. While these disparities are the products of socioeconomic factors resulting from centuries of "man's inhumanity to man" [13], exemplified in transatlantic slave trade, multi-continental colonialism, military misrule, terrorism, and other human failures, the achievable and sustainable principles of the GF can make a difference in a few decades.

It has been suggested that the Global Fund should broaden its business model to increase investments and implementation for health system strengthening (SDG Target 3.8) so as to reduce preventable deaths [5], e.g. by helping to limit health system deficiencies that contribute to epidemics, like Ebola [14], by promoting primary and secondary cancer prevention in low- and middle-income countries (SDG Target 3.d), including human papilloma virus immunization in boys and girls, hepatitis B virus immunization, cost-effective cancer screening practices, and tobacco control. Operating as the main fund-dispensing body, Global Fund can rely on technical partners like the World Bank, GAVI-The Vaccine Alliance [15–17] in tackling the infection-associated cancers of the low- and middle-income countries. This could include bringing about the unraveling of the "silent pandemic" of HTLV-I/II, which is known to be highly prevalent in these parts of the world, but where there is no evidence of control activities (see Chap. 4 in Part II). GF funding activities could also include promotion of new health technologies for cancer screening, which are applicable to all health care systems, e.g. cost-effective screening for lung cancer [18].

The Global Fund model can help meet investment needs in non-health SDG areas where proven interventions need to be scaled up with the help of innovative financing strategies [5], including public-private cooperation: e.g. small-holder farming (SDG 2), improved nutrition (SDG 2 and SDG 3), education (SDG 4), water supply and sanitation (SDG 6) and distributed rural electrification (SDG 7)

[5]. Education is a powerful instrument of change, especially for global up scaling of health if the goal of "a world converging within a generation" [2] is to be realized. In this regard, other partners that could help with global strategies in the control of cancer and HIV/AIDS in developing countries should include Global Emergency Education Fund. It has been suggested that this organization could link up with the Global Partnership in Education to form a Global Fund for Education [5], which, working on the GF model could make an impact on global standards of education and the control of cancer, HIV/AIDS, and other chronic diseases. The impact could range from improvement of health literacy within communities, and among political elites and policy makers.

"Lessons from the Global Fund can inform the work and resource mobilization of existing multilateral financing institutions, e.g., the Green Climate Fund for climate change adaptation and mitigation, the Global Environment Facility for biodiversity and ecosystem management, and the International Fund for Agricultural Development" [5]. Indeed, there is increasing realization of the fact that "health is the human face of climate change" [19], and that the health impact of climate change will be most felt in poorer countries which are least able to adapt or mitigate its effects. The Global Fund model may serve in devising adaptation and mitigation strategies that optimize health protection, especially in the low- and middle-income countries.

As the international community continues to struggle with epidemics, the most recent being the Ebola epidemic in West Africa, the need to create conditions that will facilitate vaccine development has recently been stressed [20]. The long list of human diseases for which vaccines are urgently needed, in addition to Ebola, includes HIV/AIDS, cancer, and a number of cancer-causing agents, including viruses – hepatitis type C, Epstein-Barr, various types of helminthes, and bacteria – helicobacter pylori. Cancer causing viruses for vaccines already exist are hepatitis type B virus and the human papilloma virus [21]. Figure 10.1 provides a comparison between Global Health Funds and the proposed Global Vaccine-Development Fund. The tremendous amount of money made available by these health funds is accessible almost exclusively to low- and middle-income countries. The proposed vaccine development fund would add about $2 billion to it. "The lesson we take from the Ebola crisis is that disease prevention should not be held back by lack of money at a critical juncture when a relatively modest, strategic investment could save thousands of lives and billions of dollars further down the line" [20].

Common infectious diseases are in decline around the world, thanks to increasing vaccination coverage, although the United Nations Children's Fund [22] reports that 22 million children remained unvaccinated in 2013 [23], due to a suite of socioeconomic factors as well as attitudinal beliefs. Countries failing to meet the Global Vaccine Action Plan targets, located mainly in Sub Saharan Africa and Southeast Asia, where poor vaccination coverage correlates with a lack of access to water, health care and education. Active refusal to vaccinate, however, is more likely to occur European countries [24]. It is conceivable that these broad indicators may translate in poor vaccination for the prevention of cancer causing infections.

Comparison of Existing Global Health Funds and Proposed Vaccine-Development Fund.*				
Variable	Global Fund to Fight AIDS, Tuberculosis and Malaria	GAVI	UNITAID Airline Tax	Proposed Vaccine Development Fund
Focus	HIV, tuberculosis, and malaria prevention, treatment, care, and support	Purchase and delivery of childhood vaccines	Purchase of HIV, tuberculosis, and malaria drugs	Accelerating discovery and development of new vaccines
Source of funds	Donor governments (95%); private foundations, corporate donors, and individuals (5%)	Donor governments (80%); private foundations (17%); International Finance Facility for Immunization (2%)	Airline solidarity levy	Donor governments (50%); private foundations and industry (50%) Options: financial transactions tax, tax breaks for industry donors
Eligibility	Middle- and low-income countries	Low-income countries	85% of funds must go to low-income countries	Scientists, institutions, and biotechnology companies engaged in vaccine discovery and development
Application process	Competitive country proposal	Facilitative country proposal	Funds distributed to implementing agencies and NGOs on a discretionary basis	Competitive proposal
Proposal review	Country proposals reviewed by independent technical review panel; board usually follows panel's recommendations	Country proposals facilitated by GAVI, reviewed by independent reviewers appointed by GAVI; decisions made by board	No proposals required	Proposals subject to rigorous scientific review by independent panel; board makes funding decision on the basis of scientific merit and available funds
Features	Performance-based model emphasizing results, transparency, accountability; hands-on monitoring by local fund agents and independent auditors; does not implement or fund research	Performance-based model emphasizing results, transparency, accountability; hands-off monitoring; does not implement or fund research	Does not implement or fund research	Performance-based model emphasizing results, transparency, accountability; independent auditors will monitor and assess performance; will not finance phase 3 clinical trials or conduct research
Governance	27-member international board representing donor and recipient countries, foundations, NGOs, industry, other stakeholders; 5 members are nonvoting representatives of WHO, U.N. agencies, and World Bank	28-member international board representing donor and recipient countries, private individuals, U.N. agencies, vaccine industry, foundations, other stakeholders	12-member executive board; 1 member is nonvoting WHO representative	Streamlined structure; medium-sized board whose majority of voting members represent donors; rest of composition to be determined
Funds disbursed through December 31, 2014	$25.8 billion	$7.8 billion	Approximately $2 billion	Goal: raise $2 billion initially

* Information is from the Foundation for Vaccine Research. GAVI denotes Global Alliance for Vaccines and Immunization, NGO nongovernmental organization, WHO World Health Organization, U.N. United Nations, and UNITAID Unity and AID.

Fig. 10.1 Comparison of existing global health funds and proposed vaccine development fund

10.1.1 Models of Humanitarian Global Outreach in Health Care

The core principles of the international humanitarian aid organization Médicins San Frontières (MSF) (Doctors Without Borders), including medical ethics, impartiality and neutrality, independence, bearing witness, and internal accountability [25], make the organization a natural technical partner of GF and similar funding organizations, comparable to WHO and GAVI-The Vaccine Alliance, especially since the organization has more recently realized the need to partner with both public and private entities in its work of tackling chronic epidemics, thereby leading to establishment of working relationship at national and community level [25]. This concept is what led to the creation of MSF South Africa and MSF Khayelitsha (a shanty town near Cape Town) in 2009 [26] (also see Sect. 9.14 in Part III). The new mode of collaboration is what has led to the emergence of new paradigms in HIV care, including the new concept of "the Adherence Clubs" [27], and is likely to contribute significantly in the achieving the 90-90-90 goals by 2020, and HIV/AIDS eradication by 2035.

A philanthropic initiative with the goal of "curing, preventing, and managing all diseases by the end of this century was recently launched as part of a $45 billion Chan Zuckerberg Initiative, thus, joining "forces with other philanthropists to push the envelope and support audacious ideas, with long-term commitments, to solve some of our greatest challenges… Basic research allows innovative thinkers in science and engineering to work towards ambitious and important goals, including in biomedicine" [28]. With their initial investment in this initiative, Chan and Zuckerberg would focus on the creation of a "Biohub," where engineering and life sciences research from multiple institutions in San Francisco Bay area as well as the nearby Silicon Valley would be able to interact in a "convergence" that would lead to breaking down of barriers between disciplines so as to enable beneficial cross-fertilization. Contributors to similar philanthropic endeavor include other US American billionaires, such as Paul Allen, supporting bioscience; Sean Parker, contributing to advances in cancer immunotherapy; Sandy and Joan Weill for brain research [28]. This is not to mention the tremendous commitment in multiple bio-scientific projects for which the Bill and Melinda Gates Foundation (BMGF) is renown.

The US National Institutes of Health (NIH) has a long history of global engagement in medical diplomacy grounded firmly in humanitarianism based on the moral imperative to alleviate human suffering, regardless of where it occurs. In 1940, President Franklin Roosevelt famously noted that "NIH speaks the universal language of humanitarianism…(it) has recognized no limitations imposed by international boundaries and has recognized no distinctions of race, of creed, or of color" [29]. Not only did the work of NIH scientists lay the foundation for the discovery of the retrovirus that subsequently became known as "HIV" and recognized as the causative agent of AIDS early in the pandemic (see Sect. 2.3.4.3 in Part I), the development and distribution of antiretroviral drugs to treat the retroviral infection has averted 7.6 million AIDS death in low- and middle-income countries between 2003 and 2013 [29]. The institution is in the forefront of developing new strategies to prevent and treat HIV infection, including HIV vaccine development, as well the treatment of hepatitis B virus infection, which will save several more millions of lives in the future (see Sect. 8.2.7.6 in Part III).

The Fogarty International Center of the National Institutes of Health has served for decades as one of the cornerstones of the American strategy to promote global health. Fogarty supports global health research and scientific capacity building in more than 100 countries [30].

10.1.2 The Concept of "Convergence" and the Future of Health Care Research

The future of health care research has been greatly brightened by the emergence of the concept of "convergence," which will bring about a realignment of academic structures to facilitate research and educational collaborations among biologists, clinicians, mathematicians, physicists, engineers and computer scientists [31, 32]. This would bring together great research institutions and US Government Agencies, including NIH, Department of Energy (DOE), Defense Advanced Research Projects Agency (DARPA), Food and Drug Administration (FDA), Massachusetts Institute of Technology (MIT), the David H. Koch Institute for Integrated Cancer Research, as well as others such as the National Institute of Biomedical Imaging and Bioengineering (NIBIB), the National Science Foundation [33] etc. Furthermore, many businesses are buying into the concept of "convergence". These include Verily (formerly Google Life Science), which is combining genomics with high-content and high-resolution imaging to develop individualized treatments based on biological, genetic, behavioral, and environmental data [31]. Other major companies buying into the idea include Apple, IBM, and Microsoft, which are already investing and bringing to market innovations that will allow people and institutions to track and monitor individual health, fitness and disease [31]. "Big things will happen if we can align objectives across disciplines and provide funding and programs to foster working together…creating an interagency working group on convergence with NIH, NSF, Department of Defense, FDA, and DOE, coordinated through the Office of Science and Technology Policy. In addition, a specific strategic plan for advancing biomedical science through convergence patterned on the Nanotechnology and Plant Genome initiative should be developed. NIH's Common Fund mechanism should support convergence across multiple institutes and centers" [31]. Given NIH's track record for addressing global health challenges, the great products that convergence will yield in the US will, with time, benefit the world at large, and the low- and middle-income countries in particular in the future.

10.1.3 The Concept of Global Health

The term "global health" appears to have emerged only recently, apparently coinciding with the world wide health challenges associated with the pandemic of HIV/AIDS, its causes and the responses of the global societies to them. The term certainly did not exist in spite of centuries of awareness of devastating health challenges of underdevelopment of the Third World, including high rates there of infant and maternal mortality. It also did exist when 60–70 years ago, the National Institutes of Health of the United States, through the affiliation of its National Cancer Institute with the Government of Uganda opened the Uganda Cancer Institute, primarily to study the newly described African childhood cancer Burkitt lymphoma, with its unusual association with environmental conditions and its high rate of curability by chemotherapy – at that time, a potential key disease for finding a cure to cancer.

Today, there is a wide range of definitions of what constitutes "global health". Anthropologists define it as: "(a) ethnographic studies of health inequities in political and economic contexts; (b) analysis of the impact on local worlds o the assemblages of science and technology that circulate globally; (c) interrogation, analysis, and critique of international health programs and policies; and (d) analysis of the health consequences of the reconfiguration of the social relations of international health development" [34]. Public health experts see global health as "the realities of globalization, including worldwide infectious and noninfectious public health risks" [35], while others see global health as "a goal, responding to human rights and to common interests" [36]. "What are we to make of a world with such unequal health prospects? What does justice demand in terms of global health? [37]" These were the questions in the minds of the juristically inclined.

At a recent symposium at the annual meeting of the American Society of Clinical Oncology (ASCO), the case was made for "Recognizing Global Oncology As An Academic Field", because of the fact that the majority of cancer patients are resident in the low- and middle-income countries (LMIC). "If you want to have the biggest impact in reducing cancer mortality globally, you would take the tools available in USA today (for cancer management) and use them in LMIC" [38]. At the same symposium, it was stated that "Global Health: It's not about geography, it's about perspectives" [39]. Global health concept enables appreciation of similarity and differences in health in general, and cancer in particular, across resource settings.

Whatever definition is adopted for global health, it seems that most people will agree that it is an instrument for the achievement health benefits among the people of the global communities, in spite of disparities. It enables better appreciation and understanding of health disparities in multi-ethnic populations in immigrant countries like Canada and the United States, in spite of the economically privileged positions of these nations in the world. Disparities in these countries could be due to socioeconomic, or urban vs. rural factors (see Sect. 9.15.3 in Part III). The US faces same issues in health and cancer care as occur in LMIC due to disparities in the domestic population. For example, within the United States, the incidence of cervical cancer in the southern states are higher than in the northern states, ranging from 9.2 in Texas to 9.8 per 100,000 women in Arkansas [40]. Black US women experience higher incidence and mortality than women of other races and ethnicities [40].

Thus, Global Health or Global Oncology is not necessarily a matter of "Northern Hemisphere" (often referred to as "the North") or "the developed world", and "Southern Hemisphere (also referred to as "the South) or the developing world", whereby there is usually a flow of activities from "the North" to "the South". A more functional definition should be a two-way "North – South" partnership, with a view to developing responsibilities equally based on capabilities, including research capability, creating transparency on all issues, such as publications, and promoting day-to-day application of research outcomes [41].

10.1.3.1 Models of Global Health Partnership in Cancer Control

The disparities in available health care resources are staggering. The WHO estimates that the United States and Canada have 10% of the global burden of disease, 37% of the world's health care workers, and more than 50% of the world's financial resources for health; by contrast, the African region has 24% of the global burden of disease, 3% of health workers, and less than 1% of the world's financial resources for health [42, 43]. The health resource disparity is even more extreme with cancer. In 2012, 5.3 million people died of cancer in LMIC, which exceeds the number of deaths attributed to the combination of HIV/AIDS (1.3 million), tuberculosis (1.3 million), and malaria (833,000) combined [42, 44]. In spite of this disparate relationship in the pattern of disease burden, health-related investments in funding by external agencies in LMIC are disproportionately in favor infectious diseases compared to cancer [45, 46] (also see Table 9.12 in Part III). In 2012, only 1.5% of development assistance for health (DAH) was allocated to noncommunicable diseases (NCD), including heart disease, diabetes, lung disease and cancer, of which 5% was committed specifically for cancer [47]. It has been observed that while infectious diseases continue to be a priority, "the cancer burden is greatest in the regions and environments where health care is most resource-limited and disorganized. The result is that optimal cancer care is neither available nor possible in large segments of the world" [42].

Over the last several decades, cancer management has evolved as a complex multidisciplinary practice operating in the environment of availability of costly diagnostic and treatment machinery, guided by evidence-based tools emanating from outcomes of thousands of multinational clinical trails. Practice guidelines, which have been generated by various cancer professional organizations, such as the American Society of Clinical Oncology (ASCO), European Society of Medical Oncology (ESMO), American Society of Radiation Oncology (ASTRO), and the National Cancer Institute of Canada Clinical Trials Group (NCIC-CTG). The outputs from the deliberations within these organizations are further synthesized by groups of professional experts dedicated to cancers of specific organ sites. Examples of these guideline synthesizing bodies are the British Colombia Cancer Agency (BCCA) (URL: http://www.bccancer.bc.ca) and the National Comprehensive Cancer Network (NCCN) (URL: http://www.nccn.org). The recommendations made by these bodies assume the availability of costly diagnostic and treatment resources that are standard in organized health care systems of developed countries. They make no recommendations as about how resource expenditures should be prioritized to achieve the greatest clinical benefit and outcome. This makes the applicability of existing guidelines of limited utility in LMIC [42].

10.1.3.2 The Tiered Resource Stratification Framework of the Breast Health Global Initiative (BHGI) Network

The high prevalence and mortality of cervical cancer in the LMIC, and the gross disparity in the availability of resources in high-income countries (HIC) compared to LMIC make cancer of the cervix and female breast cancer, which require multimodal therapy, model diseases for global oncology. Global socioeconomic disparities make management guidelines designed for HIC unadoptable in LMIC without appropriate modifications. This has led the Breast Health Global Initiative (BHGI) to introduce the four-tiered resource stratification framework, consisting of: [7] basic, [2] limited, [3] enhanced and [4] maximal resource levels, within which cancer management strategies can be prioritized based on available health care resources [48]. The resource stratification framework was the outcome of a series of conferences held in Seattle, Washington, USA, between 2002 and 2013 and at which BHGI experts and participants from some developing countries discussed aspects of breast cancer management, including early detection, diagnosis and treatment [49] (also see Sect. 8.2.2 in Part III). The resource-stratification methodology was first presented by BHGI in 2005 at the second BHGI Global Summit, held at the National Cancer Institute (NCI) Office of International Affairs. More than 60 international experts attended it from 33 countries of various geographic and economic developmental levels, and of diverse specialties, including radiology, pathology, surgery and radiotherapy, medical ethics, sociology and patient advocacy [48]. The BHGI resource-stratified guidelines, first published in 2006 [50], have been well received in the global health literature [51].

10.1.3.3 ASCO's Resource-Stratified Guidelines for the Management of Invasive Cervical Cancer

The American Society of Clinical Oncology (ASCO) used the BHGI guidelines as a framework to develop recommendations to clinicians and policy-makers on the management and palliative care of women diagnosed with invasive cervical cancer in different resource settings [52] in different resource settings.

The process established by ASCO included mixed method of guideline development and adaptation of the clinical practice guidelines of other organizations (e.g. Cancer Care Ontario (CCO) (Canada) [53, 54], European Society of Medical Oncology (ESMO) [55], Japan Society of Gynecologic Oncology (JSGO) [56], the National Comprehensive Cancer Network (NCCN) (USA) [57], and the World Health Organization) for the formulation of their recommendation for treatment and palliation of women with invasive cervical cancer [58]). This approach included expert consensus, which was often necessary, when there was lack of high-quality evidence-based data in extensive literature review between 1966 and 2015. Thus, when, for instance, in basic care setting, there was no literature to inform practice, the panel relied on clinical experience, training, and judgment to formulate these recommendations.

10.1.3.4 NCCN Framework for Resource Stratification for Invasive Cervical Cancer

The National Comprehensive Cancer Network (NCCN) Clinical Practice Guidelines in Oncology, which are a comprehensive set of evidence-based, consensus driven guidelines for delivering multidisciplinary cancer care across the continuum, from risk assessment through prevention, diagnosis, treatment, and survivorship, to end-of-life care, are designed for use at the resource level available in the United States [42]. As such, they are of limited utility in countries of less developed health infrastructure, as obtains in much of LMIC. However, 47% of the 754,000 verified users of the NCCN Web site are from 198 countries outside the United States and approximately 36% of the downloads of the guidelines are from outside the United States [42]. Using an adaptation of the BHGI resource-stratification guidelines, NCCN has created a formal standardized resource stratification process for its guidelines library. A description of the resource-stratification is provided in Table 10.1, ranging from basis, through limited, enhanced, to maximal. NCCN considers "basic resources" the minimum essential resources that must be available before a health care system can begin to treat a specific disease circumstance, in the absence of which successful outcome of management can not be anticipated. NCCN also advises that in the absence of basic resources, another treatment center with at least basic resources should be considered or the therapeutic focus should shift from

Table 10.1 Framework of resource stratification: primary prevention

Setting	Definition
Basic	Core resources or fundamental services that are absolutely necessary for any public health or primary health care system to function; basic-level services typically are applied in a single clinical interaction; vaccination is feasible for highest-need populations
Limited	Second-tier resources or services that are intended to produce major improvements in outcome, such as incidence and cost effectiveness, and are attainable with limited financial means and modest infrastructure; limited-level services may involve single or multiple interactions; universal public health interventions feasible for greater percentage of population than primary target group
Enhanced	Third-tier resources or services that are optional but important; enhanced-level resources should produce further improvement in outcome and increase the number and quality of options and individual choice (perhaps ability to track patients and links to registries)
Maximal	May use guidelines of high-resource settings
	High-level or state-of-the-art resources or services that may be used or available in some high-resource countries and/or may be recommended by high-resource setting guidelines that do not adapt to resource constraints but that nonetheless should be considered a lower priority than those resources or services listed in the other categories on the basis of extreme cost and/or impracticality for broad use in a resource-limited environment

"NOTE: Data adapted [50, 60]. To be useful, maximal level resources typically depend on the existence and functionality of all lower level resources." Reproduced from Ref. [59]. JGO is an Open Access publication curative treatment goals to palliative care [42]. NCCN defines "core resources" as including interventions that substantially improve outcome over those achieved with basic resources alone but are not cost-prohibitive. "Enhanced resources" are those that provide smaller incremental benefit compared to those of "core resources". They can be considered as optional when resources are particularly limited. In summary, the NCCN Guidelines represent the care recommended with 'maximal resources,' but also include all clinically appropriate choices and options, including interventions that are cost prohibitive with enhanced resources and may be aimed primarily at improving quality of life [42].

NCCN has devised its graphic resource-stratified framework in such a way that it maintains the context of the "NCCN Guidelines", which correspond to the recommendations for treatment facilities with "maximal resources". It does so by providing the information in different shadings, coloration and fonts. "This allows the users of the basic, core, and enhanced resource-versions of the NCCN Framework to immediately understand the context of the recommendations relative to care provided in the NCCN Guidelines, and to understand which therapies are optimally applied in each given resource resetting" [42]. The way this design works is best appreciated by reviewing the frameworks at the NCCN web site: http://www.nccn.org.

The resource-stratified frameworks represent a considerable effort in global health. They provide a glimpse into what is doable to achieve the best evidence-based outcomes when costs are not the limiting factors. At the same time, they demonstrate how limited available resources can be optimally used, and how treatment facilities at lower resource-settings can be upgraded as resources become more available. "The NCCN Framework for Resource Stratification provides an evidence-based approach for improving the quality, effectiveness, and efficiency of health care delivery by outlining optimal strategies for use of existing resources" [42].

10.1.3.5 ASCO Resource-Stratified Guideline for the Primary Prevention of Cervical Cancer

In order to develop an evidence-based resource-stratified guideline for primary prevention of cervical cancer globally, the American Society of Clinical Oncology (ASCO) convened a multidisciplinary discussion panel [59] and applied a mixed approach of evaluation, as discussed earlier (see Sect. 10.1.3.3), to achieve agreement on a guideline applicable to a four-tiered resource settings, ranging from basic, through limited, enhanced, to maximal, defined as shown in Table 10.1.

The recommendations of the panel are as follows [59]:

A. In all resource settings, two doses of human papillomavirus vaccine are recommended for girls aged 9–14 years, with an interval of at least 6 months and possibly up to 12–15 months. Individuals with HIV positivity should receive three doses.

B. Maximal and enhanced settings: If girls are aged ≥15 years and received their first dose before 15 years, they may complete the series; if no doses were received before age 15 years, three doses should be administered; in both scenarios, vaccination may be through age 26 years.
C. Limited and basic settings: If sufficient resources remain after vaccinating girls aged 9 to 14 years, girls who received one dose may receive additional doses between 15 and 26 years.
D. If ≥50% coverage in the priority female target population has been achieved, there are sufficient resources, and cost-effectiveness has been established, boys may be vaccinated to prevent other non-cervical human papillomavirus-related cancers and diseases.
E. Basic settings: Vaccinating boys is notrecommended.

Final Recommendations
The ASCO panel of experts underscore that health care practitioners who implement the recommendations presented in the guideline should first identify the available resources in their local and referral facilities and endeavor to provide the highest level of care possible with those resources.

The recommendations for maximal and enhanced resource settings were modified from guidelines of the WHO, Center for Disease Control (USA), and Canadian guidelines [59]. The recommendations for the limited resource setting concerning age cohort and number of doses are the same as those for the higher resource settings. Recommendations for the basic resource setting are modified from the WHO guideline [61]. Public health authorities, ministry of health, and primary care providers should vaccinate girls in the priority target age group, starting as early as possible through 14 years of age (Type of recommendation: evidence-based; evidence quality: high; strength of recommendation: strong) [59].

Cost Implications
Cost of vaccination programs appears to be the primary barrier to HPV vaccination in low-resource settings. In a series of twenty-five studies of cost-effectiveness analysis focusing on LMICs, all but one study found vaccination would be cost effect in most cases [62].

10.1.3.6 Resource-Stratified Approaches in the Secondary Prevention of Cervical Cancer

Effective cervical cancer screening by Papanicolaou (Pap) smears has resulted in the decline of cervical cancer incidence and mortality by ≥50%, mostly in high-income countries [63, 64]. Prophylactic HPV vaccination, as primary prevention of cervical cancer, is still low in LMICs, even though the vaccine has been available for more than 10 years [65, 66]. As discussed earlier, several generations of women, who are already infected with HPV are at risk of developing cervical cancer in all resource-settings, but most especially in LMICs [59]. The visual inspection with acetic acid (VIA), commonly used in screening for pre-invasive cervical lesions has been found in a large trial to reduce mortality but not incidence of cervical cancer, suggesting that VIA only served to downstage rather than prevent cancer [67]. Thus, the use VIA for cervical cancer screening can only be recommended in situations where HPV DNA testing is not practicable, such as in parts of LMICs. High-risk HPV testing has the advantage of self-collected specimens as compared to other screening methods, such as Pap smear [68], and offers advantages over provider collection, including private, at-home collection without a pelvic examination. It has been suggested that this method of screening for cervical cancer could help in expanding screening coverage in LMICs, where there is insufficient number of clinics or manpower to support nationwide clinic-based cervical cancer screening programs [63]. Not only is there ample evidence of the greater efficacy of self-collection of high-risk HPV testing over a clinic-based Pap test" [63] among unscreened/under-screened women living in LMICs [69] but also in impoverished region of a high-income country [70].

Another advantage of HPV DNA testing its potential in improving cervical cancer screening in low-resource settings, where a shortage of pathologists [71, 72], laboratories and laboratory technicians [73], colposcopists and other health care providers hinder the establishment of traditional screening program. Given the greater effectiveness of HPV DNA testing compared to traditional cytology screening test, it could be introduced in places, where less infrastructure is needed compared to what is required for cytology testing [74].

In 2014, ASCO convened an Expert Panel, as described earlier for the primary prevention of cervical cancer, to make evidence-based recommendations [75], complementing guideline development on HPV vaccination for primary prevention of cervical cancer [59], and the treatment of women with cervical cancer [52, 75]. Using the four-tiered resource-settings described in Table 10.1, the Expert Panel sought to craft guidelines for screening of women in a fashion that is appropriate for each of the resource settings, primarily for detection of pre-cancerous cervical lesions, which could be treated to avoid progression to invasive cervical cancer, and/or secondarily detect invasive cancer at an early stage, i.e. downstage the invasive cancer as may be the case in LMICs [63]. In the process, seven existing guidelines were identified and reviewed and adapted.

The recommendations of the panel are as follows [63, 76]:

A. Human papillomavirus [21] DNA testing is recommended in all resource settings.
B. Visual inspection with acetic acid may be used in basic settings, to build health care infrastructure until HPV testing becomes available.
C. Age ranges and frequencies by settings are as follows:
    (a) Maximal: ages 25–65, every 5 years;
    (b) Enhanced: ages 30–65, if two consecutive negative tests at 5-year intervals, then every 10 years
    (c) Limited: ages 30–49, every 10 years;
    (d) Basic: ages 30–49, 1–3 times per lifetime

D. TRIAGE: For basic settings, visual assessment is recommended as triage; in other settings, genotyping and/or cytology are recommended.
E. TREATMENT: For basic settings, treatment is recommended if abnormal triage results are present; in other settings, colposcopy is recommended for abnormal triage results.
F. TREATMENT OPTIONS: For basic settings, treatment options are cryotherapy or loop electrosurgical excision procedure; for other settings, loop electrosurgical excision procedure (or ablation) is recommended.
G. POST-TREATMENT FOLLOW-UP: Twelve-month post-treatment follow-up is recommended in all settings.
H. HIV INFECTION: Women who are HIV positive should be screened with HPV testing after diagnosis and screened twice as many times per lifetime as the general population.
I. POST-PARTUM SCREENING: Screening is recommended at 6 weeks postpartum in basic settings; in other settings, screening is recommended at 6 months.
J. In basic settings without mass screening, infrastructure for HPV testing, diagnosis, and treatment should be developed

A summary of the ASCO resource-stratified clinical guideline for screening women for secondary prevention of cervical cancer is provided in Table 10.2. As can be appreciated in the table, the main difference in recommendations between the various resource-settings relate to how many times a woman should e screened in a lifetime, and over what age ranges. Given the marked differences in the available management resources that are available at the resource-settings, the

Table 10.2 Summary of the ASCO resource-stratified clinical practice guideline for screening women for secondary prevention of cervical cancer

Resource strata	Method of screening	No of screens in a lifetime	Age range (years)	Management of HPV-positive patients
Maximal	HPV	9	25–64; every 5 years	Triage to colposcopy by HPV genotyping[a] and/or cervical cytology (pap)
Enhanced	HPV	5	30–64; 30, 34, 44, 54, and 64	Triage to colposcopy by HPV genotyping[a] and/or cervical cytology (Pap)
Limited	HPV	2–3 times	30–49	Triage to treatment by HPV genotyping,[a] VIA, and/or cervical cytology (Pap); VAT[b]
Basic	HPV[c]	1–3 times	30–19	VAT[b]

Reproduced with permission from Ref. [63]
Note: Adapted from www.asco.org/rs-cervical-cancer-secondary-prev-guideline
Abbreviations: *HPV* human papillomavirus; *Pap* papanicolaou, *VAT* visual assessment for treatment, *VIA* visual inspection with acetic acid
[a]HPV16 and HPV18 or HPV16, HPV18, and HPV45
[b]Determine what kind of treatment is appropriate
[c]Can start with VIA until HPV testing becomes available, permitting the development of health service delivery infrastructure

recommendations about the management of HPV-positive cases also differ. The recommendation of narrower age ranges of women at lower resource-settings is to make program development more feasible, since this limits the amount of resources expended in the screening programs. Future development of the program can be effected as resources improve over time.

10.1.3.7 Selected Global Health Activities for Promoting Cancer Control

10.1.3.7.1 Health Volunteers Overseas [77]

Health Volunteers Overseas [77], a member of the Global Health Workforce Alliance [78], is a private non-profit organization based in Washington D.C. committed to improving healthcare in developing countries through training and education, which, by emphasizing teaching, aims to create an indigenous group of trained health workers who can teach others. It has projects in Africa, Asia, Latin America, Eastern Europe, and the Caribbean and currently supports over 60 projects in more than 25 countries. It is a teaching and training organization, which engages volunteers to lecture, conduct ward rounds and demonstrate various techniques in classrooms, clinics, and operating rooms, covering a variety of specialties, including anesthesia, dermatology, internal medicine, oral and maxillofacial surgery, orthopedics, pediatrics, hand surgeons. An example of HVO activity is a Gynecologic Oncology Training Program that helps train the oncology residents that will be the direct caregivers for patients with gynecological cancer in LMICs [79].

10.1.3.7.2 Global Curriculum and Mentoring Program

Since 1985, the International Gynecologic Cancer Society [80] has been uniting the globe in the fight against gynecologic cancer by contributing to the prevention, treatment and study of gynecologic cancer, as well as improvement in the quality of life among women suffering from gynecologic cancer throughout the world [80].

In view of the fact that gynecological malignancies, especially cervical cancer, play a major role in cancer morbidity and mortality in LMICs, the International Gynecological Cancer Society [80] is launching the *Gynecologic Oncology Global Curriculum & Mentorship Program*, a comprehensive 2-year education and training program designed for regions around the world that do not currently have formal training in gynecologic oncology. The program will match institutions and individuals from higher resource settings with partners in lower resource settings wishing to obtain formal gynecologic oncology training (twinning). The specific goals of the program will be to:

- Develop and maintain a comprehensive 2-year web based curriculum for local gynecologic oncology training and education that can be adapted by each region/program to reflect local needs and facilities.

- Select trainees from lower resource settings (gynecologic oncology fellows) who will be paired with local mentor(s) from their home program as well as fully trained gynecologic oncologist(s) from a mentoring institution.
- Establish minimum requirements for the gynecologic oncology fellows to complete the program and receive a certificate following an examination and completion of the 2-year training program.
- Hold monthly tumor boards where the fellows and mentors review cases and participate in ongoing learning and mentoring opportunities.
- Establish a program for the international mentor(s) to travel to the fellow institution at regular intervals for hands-on surgical training and in-person teaching.
- Facilitate and support the fellows' travel to the mentoring institution for up to 3 months for training and education [80].

Examples of the *Global Curriculum & Mentorship Program* beginning in 2017 include:

- Da Nang General Hospital, Vietnam in collaboration with the Mayo Clinic, Jacksonville, Florida, USA and the National University Cancer Institute, Singapore
- *Hospital Central de Maputo*, Mozambique in collaboration with Barretos Cancer Hospital, Barretos, Brazil and The University of Texas MD Anderson Cancer Center, Texas, USA
- The University of the West Indies (Jamaica, Bahamas, Barbados, Trinidad & Tobago) in collaboration with the University of Miami, Florida, USA
- St. Paul's Hospital and Black Lion Hospital, Addis Ababa, Ethiopia in collaboration with the University of Minnesota and University of Michigan, USA and Martin Luther University Halle- Wittenberg, Germany
- Moi University, Moi, Kenya and the University of Toronto, Ontario, Canada

10.1.3.7.3 Global Curriculum and Mentoring Program International Development and Education Award

The International Development and Education Award (IDEA) is one of the international professional development opportunities provided by the American Society of Clinical Oncology (ASCO). It provides support for early-career oncologists in low- and middle-income countries (LMCs). The goal of IDEA is to facilitate the sharing of knowledge between oncologists in LMCs and senior ASCO members. This occurs for recipients in various ways, including: [7] receiving mentoring from a leading ASCO member; [2] attend the ASCO Annual Meetings; [3] participate in a post-Annual Meeting visit to mentors' institution; and [4] developing long-term relationships to improve cancer care in home countries [81]. Once they return to their countries, IDEA recipients are expected to share the knowledge and training they received in the program with colleagues in home institutions. The hope is that recipients will serve as agents of change to improve cancer care in their countries [81].

The Conquer Cancer Foundation, which was founded by leading members of ASCO for the achievement of their vision of a world free of the fear of cancer, works through funding of breakthrough cancer research and sharing cutting-edge knowledge with patients and physicians worldwide, and has, in the process spent more than $90 million in funding has been awarded through grants and awards to more than 65 countries since its inception [82].

ASCO, in collaboration with other organizations, offers Multidisciplinary Cancer Management Courses (MCMC) to improve cancer care globally. The course, which has been held in all continents, is designed to provide fundamental training to specialists, physicians, nurses, pathologists and oncology residents through collaboration with societies and institutions in low- and middle-income countries [83]. The first course in Africa was held January 16–18, 2007 in Abuja, Nigeria, in association with the African Organization for Research and Training in Cancer (AORTIC). In 2014, MCMCs were held in India, Armenia, Mexico, Vietnam, Brazil, and the Philippines. More than 4000 clinicians and health care workers have participated in these courses since 2004 [83].

10.1.3.7.4 Proposal for Recognizing Global Oncology as an Academic Field

At the recent (June 2–6, 2917) Annual Meeting of the American Society of Clinical Oncology, a symposium was held titled: "Recognizing Global Oncology As an Academic Field." The reason for the proposal included the fact that the majority of cancer patients are resident in LMICs, and that the problem of cancer in these countries are increasing as infectious diseases are being better controlled. Although the majority of the world population resides in LMICs, less than 5% of the money spent in oncology globally is spent there [84]. The academic field of global oncology includes activities of global organizations and agencies, challenges of international financing and development of affordable services as well as development of health systems across the globe, access to essential treatment facilities, including pharmaceutical agents, health technologies, including radiation oncology and, laboratory, medical imaging facilities and instrumentations needed for successful surgical interventions. The aim is to help in promoting access to necessary surgical care in low-income countries, where only 5% of patients, and in middle-income, where only 20% countries have access to safe surgery [84]. All in all, there is much to cover in a field of academic global oncology, and such a strategy has the potential of accelerating cancer control in LMICs.

10.2 Overcoming Barriers to Progress in the Control of Cancer and Retroviral Diseases, Including HIV/AIDS

The challenges to progress of the control of cancer and retroviral diseases in developing countries include cultural, political, educational and socio-economic factors.

10.2.1 Prioritizing Strategies for Universal Health Coverage

If the world had learnt anything at all from the devastation and massive human suffering that the Ebola crisis of 2014 caused in West Africa, it is that a decent universal health care is what is needed in the first place to avoid health care catastrophes resulting from the fragile public health infrastructure in low income countries [85]. This situation, which has been described as "the Ebola's perfect storm" [14] is relevant, not only for the control of communicable, but also non-communicable diseases, including cancer. The goals of universal health coverage (UHC) include reshaping the global health agenda [86], emphasizing that all people, irrespective of socioeconomic status, should have access to health services they need, without incurring hardship [87], thereby echoing the 1948 Universal Declaration of Human Rights, and reinforcing the importance of global health in the World Health Organization 2010 World Health Report and the 2012 United Nations General Assembly Resolution [88, 89]. As previously discussed (Sect. 9.3 and Fig. 9.1 in Part III), the unsatisfactory reception of the WHO recommendations for the adoption of UHC is particularly striking in LMICs [90] and reflects in the response of governments in LMICs to the World Health Assembly's adopted resolution 58.22 [91] concerning the control of noncommunicable diseases, including cancer (NCD) [92, 93]. While there may be various reasons for this lackluster response in, for instance, Sub-Sahara African countries, lack of political will to fund national cancer control plans has been cited in respect of countries of relatively more developed economies, such as South Africa and Nigeria, in-spite of existence there of well-conceived plans [94].

Several factors in LMICs militate against health care provision at appropriate levels. Some of these were discussed in Sect. 9.3 in Part III. Of particular concern is the national health expenditure per capita, which ranged between US$ 3966 in 2013 in Japan [95] compared to US$ 118 in Nigeria [96] and US$ 61 in India [97]. Similarly, health care insurance ranged in these countries from 99.9%, to 3% and 17% respectively. Given this pattern of national health expenditure, it is clear that provision of cancer care in Nigeria and India must be a challenging enterprise for the population in general. On the other hand, however, the challenging health provision situations in LMICs can also be conceived as opportunities for innovation in overcoming gaps in the pathway to UHC.

10.2.1.1 mHealth Strategies for Universal Health Coverage

For the past two decades, there has been a genuine revolution in worldwide telecommunication, thereby leading to nearly ubiquitous access to telecommunications technologies [98]. This has occurred under commercial pressures. The process is proving to be transformative in LMICs. Given the complexity of health care, and the promises of UHC, not only in terms of physical and mental health benefits, but also for national economies (see Chap. 9 in Part III), strategies leveraging mobile wireless technologies – mHealth – are increasingly part of a system-thinking approach to solving these challenges [99]. Global health agencies are beginning to advocate the prudent use of mHealth solutions, guided by evidence demonstrating their usability, functionality, reliability, and impact under real-world conditions [100]. WHO-led initiatives using mHealth include: mHealth Technical and Evidence Review Group (mTERG), eHealth Technical Advisory Group (eTAG), and the International Telecommunications Union-WHO Mobile Health for Non-Communicable Diseases Initiative [87]. mHealth approaches are being used to address challenges in population enumeration in Uganda [101], India and Bangladesh [87].

mHealth has the potential in ensuring adherence to evidence-based protocols with the end-effect of procuring "effective coverage" in care, i.e. the proportion of individuals receiving satisfactory health services, among those in need (target population) [102]. In Zambia, mHealth technology improves access to cervical cancer point-of-care screening; digital cervicography informs consultative telemedicine expert review determining the need for immediate referral or cyotherapy treatment [87, 103]. Mobile phone-based banking, which is expanding in sub-Saharan Africa [104] and Asia, is promoting the growth of micro-insurance and incentive schemes, leveraging technology to connect health and financing [87]. "Mobile money reduces financial barriers: It can incentivize clients to seek health care, subsidize transportation to health facilities, ensure treatment completion, and encourage savings to offset health costs. In Madagascar, Marie Stopes used mobile phone-based short message service money transfer systems to more efficiently deliver vouchers to reimburse service providers, offset client costs, and increase poor people's access to voluntary family planning services [105]. There is, thus, a need to develop mHealth strategies beyond the current vertical solutions addressing single problem focusing to integrated evidence-based packages of interoperable solutions, optimizing health systems across multiple determinants of UHC [87]. While moving from the siloed MHealth innovation to the horizontal applications required in developing health systems is bound to be challenging, it presents opportunities in addressing the complexity of global health problems of today, including those of the control of cancer and HIV/AIDS in LMICs.

10.2.1.2 Community Health Insurance Coverage as a Strategy for Universal Health Care

The importance of universal health coverage is reflected in the words of Dr. Margaret Chan, the former Director-General of the World Health Organization, who characterized it as the single most powerful concept that public health has to offer, even at times of financial crisis, when the poor are most vulnerable [106]. Much of Sub-Saharan Africa, and many other parts of the LMICs are suffering from devastating consequences of lack of access to basic health care, resulting in high infant and maternal morbidity and mortality. Dr. Babatunde Fakunle of the Centre For Sustainable Access to Health in Africa [107] provides his vision of how to bridge the critical gap in African health systems, and how governments could create the conditions for UHC. His views are based on his personal experience, working with the Shell Petroleum Development Company (SPDC), and the government of River State of Nigeria, in the creation of the Obio Community Health Insurance Scheme. The goal was to provide quality care to people who would otherwise not be able to afford it.

The Obio Community Health Insurance program is a product of the state government, SPDC and the community. The partners agreed on an annual premium of the equivalent of US$ 36.00 per participant. Every enrolled member received access to essential primary and secondary health care – with special focus on the needs of mothers and their children, including emergency obstetrics.

The Obio model, which has been endorsed by the WHO and other public health agencies, indicates that in nations where the annual capita income is between US$1000-2000, an annual investment of US$50–100 per individual could procure a meaningful amount of personal health care, thus making UHC achievable even in the poorest nations.

10.2.1.3 Innovation in Provider Training as a Strategy in UHC

Large segments of the global population lack affordable access to formally trained health care providers, especially in the rural and less developed parts of countries. In India, for example, more than half of the more than 1 billion population are typically cared for by informal providers for outpatient services [108, 109], despite regulations formally prohibiting their poor-quality practice [110], since it is inconsistent with the national health policy of "universal access to good quality of health care services without anyone having to face financial hardship" [111]. Thus, informal providers have no official role in the health care services in India, just as they are similarly excluded in in other countries, including Sri Lanka and Thailand [112]. A training intervention with 304 informal care providers in West Bengal, India, has been shown in a randomized trial to lead to improved case management, although it had no effect on inappropriate drug prescription [113]. Even though the training program lasted 72 sessions over a period of 9 months, the average attendance per session was 56%. "The training increased correct case management by 7.9

percentage points and improved adherence to disease-specific checklist by 4.1 percentage points but did not improve drug prescribing behavior" [110]. "The training in West Bengal raised correct treatment levels to that of qualified primary care providers in the public sector. But the quality remained far below acceptable levels, with 40% of the intervention group continuing to manage cases incorrectly" [110]. Thus, while better evidence of effectiveness of this innovative approach to improving access to health care is needed, the findings from the trial appears to be compelling enough to training of informal providers as a pragmatic way forward in the short term [110].

10.2.2 The Role of Politics in Attainment of Universal Health Coverage

The well-recognized knowledge-to-action gap [114], which is shared by nations in the high-income and low-income regions of the world, is related to the existence of complex industry of lobbying and knowledge translation throughout the world. It is a problem that affects the disbursement of large sums of budgetary allocation to health services, often in terms of billions of US dollars. "Thus, policy decision-makers have to find their way through evidence of varying quality and relevance, only rarely packaged to clarify the population health impact of different choices" [115]. A 2012 international forum on evidence-informed health policy-making in LMICs called for building the capacity of potential research users to evaluate and use research evidence [116]. The efficacy of this practice among legislators was the subject of an effort undertaken in Botswana [115]. Fifty-seven Botswana legislators from all parties in the parliament, with education level ranging 7 years of schooling to university degree level, agreed to participate in training organized under the auspices of the National AIDS Coordination Agency. The training, which was held on two mornings, and repeated a year later, "focused on three key areas: the value of counterfactual evidence (having a control or comparison group); how biases can distort findings ad reports; and the need for evidence about the impact on a population of policies, rather than change in individual risk…(emphasizing) the number needed to treat to prevent one adverse outcome, and the unit cost per case saved, in preference to parameters of individual impact like relative risk or simply cost of activities" [115] The participation was lively and engaging. Of 54 elected representatives, 36 attended one or both of the training sessions, including seven ministers, the deputy speaker, the leader of the opposition, and the chair of the parliamentary committee on health and HIV and AIDS. "The feedback from the Botswana legislators was very favorable; they asked for further sessions to cover the topics in more detail and for the training to be offered to other decision-makers" [115]. It is conceivable that with more training of this nature, representatives might want to have different kinds of evidence, including evidence in support of budget requests, or requesting from researchers produce about population benefits in addition to individual benefits, about the number needed treat individuals or prevent adverse outcome, as well as the unit costs per case saved of different interventions. "They might push to see the results of systematic reviews and research syntheses. Or they might at least demand studies with acceptable counterfactual evidence. They could become active and informed parties in setting the research agenda" [115].

10.2.3 The Role of Science and Engineering in Human Welfare

Several facilities that we enjoy today are the byproducts of the pursuit of knowledge for its own sake. Thus, "Guglielmo Marconi's invention of the radio was a minor additional step after James Maxwell's formulation of the law of electricity and magnetism and Heinrich Hertz's detection of electromagnetic waves. Likewise, the human world would be helpless without electricity, yet when Michael Faraday enabled its use by discovering the induction of electric current, his goal was to understand the universe, not utility" [117]. Other byproducts of basic science include the detection of blood cells and the entire field of bacteriology [118]. A deep appreciation of basic science as the foundation of human progress in all facets of life, including economics, engineering, and health, is much needed in the world in general, and in LMICs in particular.

10.2.3.1 The Role of Science

Science has been described as "an amazing human invention – a huge community effort to discover truth through repeated cycles of testing and self-correction" [119]. For humanity to prosper, individuals and communities need to be able to act and make decisions based on the best available evidence, rather than "in an emotion-based manner that is strongly influenced by the beliefs of their cultural cohort" [119]. As discussed in Sect. 1.8 in Part I, examples the failure of the appreciation of the role of science in governance might have been responsible for the delay in addressing the HIV/AIDS crisis in southern Africa [120, 121]. The pervasive lack of healthy relationship to science in LMICs continues to bedevil the control of deadly diseases, including cancer and HIV/AIDS [122]. Unfortunately, this is a global problem, including populations of countries. In a recent survey, only 32% of US adults supported a statement that "climate scientists' research findings are influenced by the best available scientific evidence most of the time" [119]. The gap in scientific education in scientific education has to be a major obstacle in cancer and HIV/AIDS control in LMICs.

10.2.3.1.1 Science Advocacy

Science advocacy involves the work of a person or group of persons communicating the nature of science and its potential of bringing about changes within the society. The best people to do the work are scientists. Being outstanding in one's field does not necessarily make one a good communicator. The ideal scientific advocate is also someone who is able to communicate the message. "So scientists should recognize their star quality and hone their skills to become effective advocates for science, whether at elementary schools, museums, churches, or even the halls of government" [123].

"Narrative deficiency" has been identified as the single biggest problem facing science today [124]. Science should be communicated like a story, using the "And, But, Therefore" or "ABT" method. "A scientist could say, for example, 'I can tell you that in my laboratory we study physiology AND biochemistry, BUT in recent years we've realized the important questions are at the molecular level, THEREFORE we are now investigating the following molecular questions…." [124, 125]. This approach improves the delivery of the scientific message, "…when crafting a scientific narrative, it us important to proceed with the same caution and precision as one would approach a scientific experiment…" [125].

10.2.3.1.2 Science Education in Low- and Middle-Income Countries

The teaching of science is well grounded in LMICs countries, where major languages of the world, e.g. Arabic, Chinese, Iranian, etc., are used in education. In several countries of Sub-Saharan Africa, the languages of commerce and governance are those of the former colonial powers, e.g., French, English, Spanish, Portuguese, etc. Following the examples of India and China, Nigeria recently announced a government project to commence nationwide teaching of mathematics and science subjects in indigenous languages. "For us to build the country of our dreams, for us to make Nigeria a truly great nation, a nation that is able to feed and house its citizens, a nation with a stable currency, we must embrace science and technology…Nigeria will remain dependent nation if the citizens did not embrace science and technology because no nation can become great without science and technology, the Minister of Science and Technology is quoted as announcing recently [126, 127].

10.2.3.1.3 Examples of the Transforming Effect of Science in LMICs

As countries of limited resources embrace science and technology, examples of dramatic achievements are emerging, which are impacting positively on other countries in the category.

10.2.3.1.3.1 Impact of Science on the Emergence of South Korea

About 50 years ago, South Korea was one of the poorest countries in the world, recovering from the Korean War and facing serious economic difficulties. Today, it is a member of the G20 major global economies with a market value that ranks 11th in the world. This miraculous growth has been attributed to investment in science and technology, behind which stands the Korea Institute of Science and Technology [128], which recently marked its 50th anniversary [129]. The success of South Korea has been attributed to its visionary leader, President Park Chung-hee, who recognized early the role of science and technology in nation building, and promoted the emergence of vibrant culture of research and development (R&D). "Given that KIST was born out of international development aid, particularly from the United States, the institute strives to serve as a model to help other countries and is currently assisting Vietnam in establishing a research institute, the Vietnam-Korea Institute of Science and Technology…Over the next 50 years, all nations will need to make great advances in meeting the challenges of a growing population, sustainable energy and food resources, health and disease issues, and climate change" [129].

10.2.3.1.3.2 Impact of Science on Middle-Eastern Cooperation

A scientific collaboration involving Cyprus, Egypt, Iran, Israel, Jordan, Pakistan and the Palestinian Authority, which recently led to the inauguration of the Synchrotron-light for Experimental Science and applications in the Middle East (SESAME) represents the power of science in bringing together countries under a common goal of advancing knowledge for the benefit of all mankind [130]. "The light source can elucidate the structure and properties of matter, with applications in physics, material science, biology, and medical imaging" [130]. The success of this project is already stimulating interest in Sub-Saharan Africa, which has proposed the development of the African Light Source and the Light Sources for Africa, the Americas and Middle East Project (LAAMP). "These new endeavors will face challenges. But they share with SESAME the goals of building regional capacity and promoting understanding, friendship, and peace by bringing together scientists from different countries and ethnicities to perform world-class science" [130].

10.2.3.1.3.3 Impact of the US-Cuba Scientific Diplomacy

The century old US-Cuban relationship in science [131], which was chilled for about a half-century of estrangement, has been jolted back to life by the recent political détente. This relationship is an example of how collaboration between very different political and cultural entities can benefit human at large and has the potential of a boon to LMIC science. Recently, "the American Association for the Advancement of Science (AAAS) and the Cuban Academy agreed to jointly focus on biomedical research in cancer, infectious diseases, drug resistance, and neuroscience, and earlier this year (2015), the two countries discussed how to start working together to protect the marine environment between them – a mere 90 miles of ocean, where among other shared challenges, both faced the Deepwater Horizon catastrophe" [131].

The history of the development of science and technology is one that most LMICs could learn from. Carlos Finlay, the scientist who first hypothesized that mosquitoes transmitted yellow fever, is credited as being the father of Cuban science [132]. His key insight into yellow fever occurred in 1881, when Cuba was still a colony of Spain. Intellectualism, however, withered following independence in 1898, as the island nation became the winter playground for the rich and famous Americans [133]. By 1950, the field of applied research was dead in Cuba, and laboratory activities were moribund. Cuba's revolution of 1959 changed all that, as Fidel Castro, in a speech in January 1960 declared: "The future of our country has to be necessarily a future of men of science, of men of thought" [133]. By that time, 20% of Cubans were illiterates, and the leader's vision was like a pipe dream. Things, however, began to change, when in 1965 the government established the National Center for Scientific Research, and research centers were created at various parts of the island for various fields of endeavor, including metallurgy and sugarcane byproducts [133]. "In 1973, the first Ph.D. was awarded in Cuba, in neuroscience. Cuba now has 63 universities, and roughly one scientist for every 1800 people" [133].

By 1981, Fidel Castro had begun work in creating Biotech in Cuba, based partly on the development of interferon as a weapon against cancer. "Cuba's state-owned biotech industry now employs more than 21,000 people at 32 institutes and enterprises managed by BioCubaFarma, a holding company. Biotech is now the country's biggest source of revenue, after tourism, earning several hundred million dollars each year from exports of products such as recombinant epidermal growth factor for diabetic ulcers; recombinant erythropoietin for anemia; and a pentavalent vaccine against diphtheria, tetanus, whooping cough, hepatitis B, and *Haemophilus influenza* B" [133].

The country's science and technology has suffered from the economic embargo imposed by its economic giant neighbor, the United States, as well as from the vagaries of "the special period," as the period of existential crisis resulting from the collapse of the Soviet Union is referred to. In spite of its virtual isolation from the cyber world, in which the identification of the Cuban IP address during a software downloading process induces the response "You are in a forbidden country," Cuban scientists and engineers have been able to their country's first computer from the scratch [133].

10.2.3.1.3.4 Boosting Science in Singapore

The government of Singapore recently announced an investment of 19 billion Singapore dollars (US$13.2 billion) on research and development for the period 2016–2020, representing an 18% increase over the previous 5-year cycle. The biggest share of the budget – 21% – has been earmarked for health and biomedical sciences. "The budget also boosts A*STAR's high profile Biopolis research complex, and places a priority on advanced manufacturing technologies, with emphasis on boosting the aerospace electronics, chemical, pharmaceutical, and marine sectors [134].

10.2.3.1.3.5 India Boosts its Ambitious Space Program

India has an ambitious space program with a plan to put a lander on Mars in 2021 or 2022 and send an orbiter to Venus shortly thereafter. This is a progression of its Mars Orbiter Mission in which has led to successful entry of an orbit around the Red Planet in 2014, "and continues to beam back stunning images of the planet to Earth" [135]. The country is also planning a moon mission early in 2018, which will debut a rover. The Indian Space Research Organization (ISRO) is also contemplating a Venus mission, with the aim of studying the planet's carbon dioxide-rich atmosphere, with a view to understanding its influence in the buildup of greenhouse gas in our planet's atmosphere.

10.2.3.2 Transforming the Practice of Engineering to Meet Global Health Needs

"More of the world's population has access to cell phones than to basic sanitation facilities, a gap that can only be closed if the engineering and international aid communities adopt new approaches to design for scarcity and scalability" [136]. About 50% of people who lack access to water and sanitation, among 789 million people with no access to safe drinking water, 2.5 billion with no basic sanitation and 1 billion practicing open defecation, live in middle-income countries [137]. Technologies that have shaped the public health systems of the wealthy countries have proved difficult to export to the developing world, because they were built at high cost to solve problems of high-resource settings, and are not adaptable as infrastructures of poor countries. Health technologies that are used in developed countries for the control of cancer and HIV/AIDS often fail in low-resource settings because of maintenance and repair challenges, lack of technical support, as well as intermittent power supply that render equipment unusable, often in spite of availability of donor funds [138, 139]. "If we are to resolve global inequalities in access to innovations that improve health, we must adopt new approaches to engineering design that reflect the unique needs and constraints of low-resource settings.

The need to innovate engineering solutions for social constraints of low-resource settings leads to "frugal designs" [136]. Thus, when reusing of disposable syringes within the setting of "cultural resistance to waste" led to a high rate of infectious complications in injection recipient, a solution was found in the design of the non-reusable syringe. Fragile technologies require understanding which features are most important to support positive impact at scale, while maintaining efficacy and a user-centric focus [136]. "Ensuring long-term access to new health technologies requires a coordinated architecture that integrates efforts to make new technologies affordable, makes certain the technologies are available where they are needed, and facilitates adoption of the technologies within health systems" [136, 140].

Engineering students in LMIC must be educated to become successful practitioners of frugal design from a system perspective [141]. Curricular reforms are needed in LMICs' engineering schools to change "learning foci that are too theoretical, based on outdated curricula, and not relevant to local needs. The teaching and learning approaches that emphasize role memorization stunt students' potential to be innovative" [136].

There is a need in rethinking approaches to the implementation of public infrastructures needed for health and disease control. To explain why more people have access to mobile phones than toilets, it is because "adoption (of a new technology) is facilitated when end-users see a direct personal benefit associated with purchase of a new technology. Access to mobile phones increased profits to fishermen in India and market participation for farmers in Uganda. In contrast, the benefits of health or sanitation technologies may not be as apparent to end-users" [136, 142].

"It is time for the engineering and international aid communities to adopt approaches that can improve global health in ways that can be sustained" [136].

10.2.3.3 Models in Innovation with Potentials for Cancer and HIV/AIDS Control in LMICs

The following are selected examples of measures in LMICs, which are responding to some of the barriers for effective health care, including cancer and HIV/AIDS control.

10.2.3.3.1 Affordable, Weather Durable X-ray Machine

Some the challenges of availability of diagnostic equipment in developing countries include equipment design for operations in weather conditions in tropical areas. GlobalDiagnostiX of the Swiss Federal Institute of Technology in Lausanne recently unveiled its new x-ray machine prototype that has been developed to address some weather-related frailty of x-ray machines. The machine is apparently designed to weather heat, dust, and humidity and an unreliable electricity supply, at a quoted cost of about one-tenth of the cost of current equipment [143].

10.2.3.3.2 Affordable Cancer Immunotherapy in Cuba

The Center of Molecular Immunology (CIM) in Havana, Cuba, which has been in cancer vaccine business for several years, is believed to be providing cancer vaccines for use in the country [144]. In a trial run, CIM in 2014 began stocking 50 primary care units across the country with two lung cancer vaccines. Vaccines against breast cancer and other tumor types are also being developed there [144].

10.2.3.3.3 The Smart Village for Rural Electrification

Recognizing that "energy is a catalyst for improving the quality of life, the UN Sustainable Development Goals (SDGs), which succeeded the Millennium Development Goals (MDGs) in 2015, and would run for 15 years, has identified Goal 7 among its 17 goals: it aims to ensure "access to affordable, reliable, sustainable and modern energy for all" [145]. This would be a potential solution to the energy gap affecting an estimated 1.3 billion people who have no access to electricity, including 70% of the world's poor people who live off-grid in the countryside. Several countries in LMICs struggle to provide constant electricity supply even to their densely populated cities, with significant disruption of critical health care services, including those required to cancer and HIV/AIDS control. This is because developing electrical grips to comprehensive electricity supply is hampered by several impediments, including prohibitive expense, even for a petroleum based economy such as that of Nigeria [146].

Access to sustainable energy is a necessary precursor to all the other SDGs and its acquisition calls innovative strategies. "One such strategy is the 'smart village,' a rural analog of the 'smart city' concept, together with modern information and communication technologies, enables holistic development, including cultural changes in the provision of good education and health care; access to clean water, sanitation, nutrition; and the growth of social and industrial enterprises to boost income" [145]. This idea is gradually taking root in parts of LMICs, including northern Tanzania. In Bangladesh and Rwanda, the "smart village concept" is manifesting in form of solar home systems and mini-grids, respectively [145]. "In the 15-year SDG marathon that lies ahead, something more than warm words and sentiments of Goal 7 must be done to cover the "last mile": the millions of people living in remote communities with no realistic chance of being connected to a grid" [145].

10.2.3.3.4 HIV/AIDS Control in China

The history of how China transformed its response to HIV/AIDS pandemic, after its relatively late arrival in the country is valuable to the rest of LMICs [147]. China was at the beginning of the pandemic "HIV laggard," because it was felt that the "absence" of the risk factors and the "social evils" for HIV/AIDS would protect the country from the pandemic. However, when the role of plasma sellers operating in the rural areas was finally recognized, the response of the government has made it a global leader in control of HIV/AIDS [148].

The government of China responded to HIV/AIDS with significant investment, rising from US$2 million in 2000 to US$600 million in 2015, "health officials reduced the number of steps between HIV screening and treatment from 4 to 1, thus increasing the percentage of individuals with a confirmed HIV infection who had initiated antiretroviral therapy from 40% to 90%," as well as offering free antiretroviral therapy to HIV positive individuals, regardless of their CD4 cell count [148].

This is precisely what the Joint United Nations Programme on HIV/AIDS hopes would contribute to the 90-90-90 targets, as discussed in Sect. 8.2.7.7 in Part III. "In particular, countries dependent on donor funds may want to take note of the Chinese government's insistence on using international assistance to meet national objectives rather than donor's objectives and on satisfying grantors' reporting requirements with a single, centralized data platform built to meet national needs and reduce paperwork" [148].

10.3 The Future of Control of Cancer and Retroviral Diseases

The last five decades have witnessed so much progress in both cancer and HIV/AIDS control that one can confidently say that the end the fearsomeness of the diagnosis of these diseases is in sight. The aim of this section is to review some of the achievements that have made this assertion tenable.

10.3.1 From the National Cancer Act to the Twenty-First Century Cures Act

In December 1971, the "War Against Cancer" was set in motion by the signing to law the "National Cancer Act" by Richard Nixon, a US Republican President [149]. The National Cancer Act was meant to bring "an end" to cancer, a Republican equivalent to the Democrats' moon landing a decade earlier. Forty-Five years later, the signing to law of the "21st Century Cures Act" in December 2016 by Barack Obama, a US Democratic President, signaled the end of that war [150]. The Cures Act was based an initiative of Vice President Joe Biden's Cancer Moonshot, who had lost his son Beau to brain cancer in July 2015, designed to accelerate the pace of cancer drug development [151]. While the declaration of the war against cancer of 1971 was based largely on emotion, as so little was known at the time about the enemy, cancer, (see Sects. 2.1 and 2.1.1 in Part I), the initiative of the Cancer Moonshot was build on the massive amount of knowledge that had accrued from billions of US dollars in cancer research investment over the preceding several decades. The Moonshot initiative was subsequently worked on by the Blue Ribbon Panel (BRP) of scientific experts convened to make recommendations to the National Cancer Advisory Board, the adviser to the National Cancer Institute (NCI) on research opportunities for acceleration [152].

Shedding more light of the BRP, its 3 co-chairs describe the working of the 150 people, including scientists, clinicians, patients' advocates and industry representatives worked to identify critical areas of cancer research promised by the Cancer Moonshot [152]. The NCI also led a campaign to collet input from the wider research community and the public. "Thus, the recommendations of the BRP reflect what the broader community sees as ripe for progress" [152]. The recommendations highlight the need and desire for the participation of the general public in advancing the goals of the Moonshot, not only in the treatment of cancer but also in its preventions, by identification of the genetic basis of cancer and identification for predisposition in the general public and early initiation of preventive measures. Some of the recommendations will address the heterogeneity in treatment response to standard therapy and how treatment outcomes could be predicted in the future. Thus, there is a recommendation for a pilot project in which all patients with newly diagnosed colorectal or endometrial cancer would be screened for DNA mismatch-repair deficiency; those with such defects would then have targeted genome sequencing for mutations in DNA mismatch-repair genes associated with Lynch syndrome, so that the discovery of such genetic abnormality would inform ways of identification of members of the family for potential prevention of morbidity and mortality from diseases. A better understanding of childhood cancer and the role of fusion oncoproteins in the etiology of pediatric cancers, as well as the failure of immunotherapy in children is also the subject of a recommendation.

The panel's suggestions, which basically are "for making a decade's worth of progress in cancer advances over the next five years" include [153]:

1. Establish a network for direct patient engagement to provide patients with a genetic profile of their cancer and allow them to "preregister" for clinical trials, so they can be contacted when an eligible trial opens.
2. Create a clinical trials network devote exclusively to immunotherapy to help scientists understand why this promising is effective only in some patients. Establishing clinical trials networks devoted exclusively to immunotherapy for pediatrics and adult cancers could lead to new vaccines to prevent cancers of all types in children and adults.
3. Identify therapeutic targets to overcome drug resistance through studies that determine the mechanisms leading cancer cells to become resistant to previously effective treatments.
4. Build a national cancer data ecosystem for sharing and analyzing data, so researchers, clinicians, and patients will be able to contribute data to help facilitate efficient data analysis.
5. Intensify research on the major drivers of childhood cancers to improve researchers' understanding of fusion oncoproteins in pediatric cancers and use new preclinical models to develop inhibitors that target them.
6. Accelerate the development of guidelines for the routine monitoring and management of patient-reported symptoms to minimize debilitating side effects of cancer and its treatment and to help patients stay on their drug regimens and improve their quality of life.
7. Reduce cancer risk and cancer health disparities through approaches in development, testing, and broad adoption of proven prevention strategies. This recommendation also calls increasing testing for hereditary cancer syndromes in

people with certain types of cancer and their family members so that identified as at high risk can begin early prevention or screening efforts.
8. Mine past patient data to predict future patient outcomes.
9. Create a three-dimensional cancer atlas of genetic lesions and cellular interactions I tumor, immune, and other cells in the microenvironment that maps the evolution of tumors from development to metastasis to enable researchers to develop predictive models of tumor progression and response to treatment.
10. Develop new cancer technologies to characterize tumors and test therapies, including implantable microdosing devices (which deliver drugs directly into a tumor), to test their effectiveness and advanced imaging technologies to study cancers at extremely high resolution, to deliver smarter, more effective therapies to patients.

10.3.2 Advances in Lung Cancer Prevention

Lung cancer is a major cause of morbidity and mortality in high-income countries as well as LMICs (see Sect. 7.4.1 in Part II). In Ontario, Canada, lung cancer is responsible for the largest cancer death (49.9 per 100,000 individuals), despite falling smoking rates [154–156]. The implementation of a lung cancer-screening program, in addition to the continued efforts in primary prevention of smoking, could reduce lung cancer mortality. However, concerns have been raised about cost-effectiveness of such a cancer prevention program [157, 158]. In the US, lung cancer screening is recommended for current and former smokers who have quit within the past 15 years, aged 55 through 80, who smoked at least 30 pack-years [18, 159]. Despite lung cancer screening being recommended by a number of organizations, the cost-effectiveness of lung cancer screening has remained uncertain [160–164].

The cost-effectiveness of lung cancer screening is related to screening policy characteristics [18]. A micro-simulation model was used to analyze 576 different lung cancer-screening policies for persons born between 1940 and 1969 in Ontario, Canada found that requiring stringent smoking history eligibility criteria (i.e. requiring higher levels of accumulated smoking exposure) was more cost-effective than less stringent smoking history eligibility criteria. Thus, limiting screening to individuals with substantial past smoking histories may allow lung cancer screening to be implemented in cost-effective manner. Annual screening is also suggested to be more cost-effective than biennial screening [18]. The global applicability of this lung cancer control strategy would depend on local cost-effectiveness analysis.

10.3.3 Ending the HIV/AIDS Pandemic

The beginning of the end of HIV/AIDS began at the HIV/AIDS international conference of July1996, when the world learnt of the effectiveness of a new regimen of antiretroviral therapy (ART), with a protease inhibitor as the centerpiece of the regimen. It is estimated that between 2000 and 2014, ART had saved an estimated 7.7 million worldwide [165]. The availability of this new weapon led to a series of investigations with a view of defining in the continuum of the HIV/AIDS how soon ART should be used, ultimately leading to the establishment of the concept of pre-exposure prophylaxis in HIV-negative individuals engaged in high-risk sexual activities [166]. Subsequent studies, including the START study [167], the TEMPRANO [7], and the HPTN 052 study [168] provide evident-based approaches for the effective treatment and prevention of HIV infection (also see Sect. 9.13 in Part III). When this evidence-based blue-print is linked with UNAIDS 90-90-90 goal for 2020 to end the HIV/AIDS epidemic by 2030 (see Sect. 9.14 in Part III), it becomes clear that what is now required to put an end to HIV/AIDS is a final push in collaboration between national governments, funding organizations and the appropriate political commitment [169].

10.3.4 Finding a Cure for AIDS – A Possible Convergence of Cancer and HIV/AIDS Research

ART has the potential to abort new HIV infection, thereby benefitting both the individual and communities. However, once the infection occurs, ART is unable to get rid of the virus in the body. ART is not curative. This is because of the life cycle of HIV, as a retrovirus, integrating into the host genome, and thereby retaining the capability to replicate for many years [170–173], starting from the stage of acute infection [174–176], thus, leading to lifelong infection and rebounds of the infection in most infected individuals following discontinuation of ART [176, 177]. The viral reserve and the sanctuary sites are the subjects of ongoing research [178].

The establishment and maintenance of the viral reservoir appears to be affected at least in part by the immune system, thus, raising the prospect of immunotherapy, the elucidation of the mechanism of which is proving revolutionary in cancer medicine, and, perhaps, signaling a potential convergence of cancer and HIV/AIDS treatment strategies [170].

References

1. WHO. The world health report 2000: health systems: improving performance. Geneva: World Health Organization; 2000.
2. Jamison DT, Summers LH, Alleyne G, Arrow KJ, Berkley S, Binagwaho A, et al. Global health 2035: a world converging within a generation. Lancet. 2013;382(9908):1898–955.

3. Beaglehole R, Bonita R, Alleyne G, Horton R, Li L, Lincoln P, et al. UN high-level meeting on non-communicable diseases: addressing four questions. Lancet. 2011;378(9789):449–55.
4. IHME. Institute for Health Metrics and Evaluation, development assistance for health database 1990–2015 (IHME, Seattle, WA, 2016). Seattle: Institute for Health Metrics and Evaluation, Development Assistance for Health Database; 2015. Available from: http://ghdx.healthdata.org/record/development-assistance-health-database-1990-2015
5. Sachs JD, Schmidt-Traub G. Global fund lessons for sustainable development goals. Science. 2017;356(6333):32–3.
6. GF. GF, "The right side of the tipping point for AIDS, tuberculosis and malaria: investment case for the Gobal Fund's 217–2019 replenishment". Geneva: Global Fund; 2015.
7. Group TAS. A trial of early antiretrovirals and isoniazid preventive therapy in Africa. N Engl J Med. 2015;2015(373):808–22.
8. Bornemisza O, Bridge J, Olszak-Olszewski M, Sakvarelidze G, Lazarus J. Health aid governance in fragile states: the Global Fund experience. Glob Health Gov. 2010;4(1)
9. Nahlen BL, Low-Beer D. Building to collective impact: the Global Fund support for measuring reduction in the burden of malaria. Am J Trop Med Hyg. 2007;77(6 Suppl):321–7.
10. WHO. WHO releases new guidance on insecticide-treated mosquito nets (press release). Geneva: World Health Organization; 2007. Available from: http://www.who.int/mediacentre/news/releases/2007/pr43/en/
11. Samb B, Desai N, Nishtar S, Mendis S, Bekedam H, Wright A, et al. Prevention and management of chronic disease: a litmus test for health-systems strengthening in low-income and middle-income countries. Lancet. 2010;376(9754):1785–97.
12. Sherry J, Mookherji S, Ryan L. The five-year evaluation of the Global Fund to fight AIDS, tuberculosis, and malaria: synthesis of study areas 1, 2 and 3. Geneva: GFATM and Macro International; 2009.
13. Smith L. Man's inhumanity to man'and other platitudes of avoidance and misrecognition: an analysis of visitor responses to exhibitions marking the 1807 bicentenary. Mus Soc. 2010;8(3):193–214.
14. Piot P. Ebola's perfect storm. Science. 2014;345(6202):1221.
15. GAVI. GAVI Alliance: delivering on the promise. 2016. http://www.gavi.org/Library/Publications/Pledging-conference-for-immunisation/Delivering-on-the-promise/
16. GAVI-Alliance. GAVI Alliance. Human pappilomavirus vaccine support. Oct 16, 2013. 2013. Available from: http://www.gavialliance.org/support/nvs/human-papillomavirus-vaccine-support. Accessed 22 Apr 2017.
17. Hadler SC, Fuqiang C, Averhoff F, Taylor T, Fuzhen W, Li L, et al. The impact of hepatitis B vaccine in China and in the China GAVI project. Vaccine. 2013;31:J66–72.
18. ten Haaf K, Tammemägi MC, Bondy SJ, van der Aalst CM, Gu S, McGregor SE, et al. Performance and cost-effectiveness of computed tomography lung cancer screening scenarios in a population-based setting: a microsimulation modeling analysis in Ontario. Can PLoS Med. 2017;14(2):e1002225.
19. Hunter DJ, Frumkin H, Jha A. Preventive medicine for the planet and its peoples. N Engl J Med. 2017;376(17):1605–7.
20. Plotkin SA, Mahmoud AA, Farrar J. Establishing a global vaccine-development fund. N Engl J Med. 2015;373(4):297–300.
21. Clifford GM, Goncalves MAG, Franceschi S. HPV, group Hs. Human papillomavirus types among women infected with HIV: a meta-analysis. AIDS. 2006;20(18):2337–44.
22. UNICEF. State of the world's children: celebrating 20 years of the Convention on the Rights of the Child. New York: Unicef; 2009.
23. CA. Mapping the barriers to vaccination. Science. 2016;353(6307):1511.
24. de Figueiredo A, Johnston IG, Smith DM, Agarwal S, Larson HJ, Jones NS. Forecasted trends in vaccination coverage and correlations with socioeconomic factors: a global time-series analysis over 30 years. Lancet Glob Health. 2016;4(10):e726–e35.
25. Chen K. To run where the brave dare not go. Science. 2014;346(6215):1304.

26. Fox RC. Doctors without borders: humanitarian quests, impossible dreams of Médecins Sans Frontières. Baltimore: JHU Press; 2014.
27. Campion EW. Treating millions for HIV—the adherence clubs of Khayelitsha. N Engl J Med. 2015;372(4):301–3.
28. Baltimore D. The boldness of philanthropists. Science. 2016;10(1126):1473.
29. Fauci AS, Collins FS. Benefits and risks of influenza research: lessons learned. Science. 2012;336(6088):1522–3.
30. NIH. Fogarty International Center, Fogarty's role in global health. 2017. Available from: http://www.fic.nih.gov/About/Pages/role-global-health.aspx
31. Sharp P, Jacks T, Hockfield S. Capitalizing on convergence for health care. Science. 2016;352(6293):1522–3.
32. Sharp P, Hockfield S. Convergence: the future of health. Science. 2017;355(6325):589.
33. Coates RJ, Bransfield DD, Wesley M, Hankey B, Eley JW, Greenberg RS, et al. Differences between black and white women with breast cancer in time from symptom recognition to medical consultation. Black/white cancer survival study group. J Natl Cancer Inst. 1992;84(12):938–50.
34. Janes CR, Corbett KK. Anthropology and global health. Annu Rev Anthropol. 2009;38:167–83.
35. De Cock KM, Simone PM, Davison V, Slutsker L. The new global health. Emerg Infect Dis. 2013;19(8):1192–7.
36. Berlinguer G. Globalization and global health. Int J Health Serv. 1999;29(3):579–95.
37. Dwyer J. Global health and justice. Bioethics. 2005;19(5–6):460–75.
38. Schulman L. Recognizing global oncology as an academic field. Annual Conference of the American Society of Clinical Oncology. Chicago, IL; 2017 (Personal communication).
39. Trimble E. Global Health: it's not about geography, it's about perspectives. Annual Conference of the American Society of Clinical Oncology, Chicago, IL; 2017 (Personal communication).
40. Howlader N, Noone A, Krapcho M, Garshell J, Miller D, Altekruse S, et al. SEER Cancer Statistics Review, 1975–2012, National Cancer Institute. Bethesda. 2015. http://seer.cancer.gov/csr/1975_2012
41. Alatise OI. Key principles to strengthening partnerships. Meeting of the African Organisation for Research and Training in Cancer, North America Region, at the Annual Conference of the American Society of Clinical Oncology, Hilton Chicago, June 3rd, 2017 (personal communication).
42. Carlson RW, Scavone JL, Koh W-J, McClure JS, Greer BE, Kumar R, et al. NCCN framework for resource stratification: a framework for providing and improving global quality oncology care. J Natl Compr Cancer Netw. 2016;14(8):961–9.
43. WHO. The world health report: working together for health. 2006. Available from: http://www.who.int/whr/2006/en/
44. Beyene T, Hoek H, Zhang Y, Vos T. Global, regional, and national age–sex specific all-cause and cause-specific mortality for 240 causes of death, 1990–2013. Lancet. 2015;385(9963):117–71.
45. Fromer MJ. Cancer care in low-resource areas: some improvements over the years, but serious problems remain. The ASCO Post. December 10, 2015.
46. Abubakar I, Tillmann T, Banerjee A. Global, regional, and national age-sex specific all-cause and cause-specific mortality for 240 causes of death, 1990–2013: a systematic analysis for the Global Burden of Disease Study 2013. Lancet. 2015;385(9963):117–71.
47. Nugent R, Feigl A. Where have all the donors gone? Scarce donor funding for non-communicable diseases. Washington, DC: Center for Global Development; 2010.
48. Anderson BO, Duggan C. Resource-stratified guidelines for cancer management: correction and commentary. J Glob Oncol. 2016;3:84–8. JGO006213
49. Anderson BO, Braun S, Carlson RW, Gralow JR, Lagios MD, Lehman C, et al. Overview of breast health care guidelines for countries with limited resources. Breast J. 2003;9(s2):S42–50.
50. Anderson BO, Shyyan R, Eniu A, Smith RA, Yip CH, Bese NS, et al. Breast cancer in limited-resource countries: an overview of the breast health global initiative 2005 guidelines. Breast J. 2006;12(s1):S3–S15.

51. Echavarria MI, Anderson BO, Duggan C, Thompson B. Global uptake of BHGI guidelines for breast cancer. Lancet Oncol. 2014;15(13):1421–3.
52. Chuang LT, Temin S, Camacho R, Dueñas-Gonzalez A, Feldman S, Gultekin M, et al. Management and care of women with invasive cervical cancer: American Society of Clinical Oncology resource-stratified clinical practice guideline. J Glob Oncol. 2016;2(5):311–40.
53. Hirte H, Kennedy E, Elit L, Fung MFK. Systemic therapy for recurrent, persistent, or metastatic cervical cancer: a clinical practice guideline. Curr Oncol. 2015;22(3):211–9.
54. Elit L, Kennedy E, Fyles A, et al. Follow-up for cervical cancer. 2015. Available from: https://www.cancercare.on.ca/common/pages/userFile.aspx?fileId=340742
55. Colombo N, Carinelli S, Colombo A, Marini C, Rollo D, Sessa C, et al. Cervical cancer: ESMO clinical practice guidelines for diagnosis, treatment and follow-up. Ann Oncol. 2012;23(Suppl 7):vii27–32.
56. Ebina Y, Yaegashi N, Katabuchi H, Nagase S, Udagawa Y, Hachisuga T, et al. Japan Society of Gynecologic Oncology guidelines 2011 for the treatment of uterine cervical cancer. Int J Clin Oncol. 2015;20(2):240–8.
57. Koh W-J, Greer BE, Abu-Rustum NR, Apte SM, Campos SM, Chan J, et al. Cervical cancer. J Natl Compr Cancer Netw. 2013;11(3):320–43.
58. WHO. Comprehensive cervical cancer control: a guide to essential practice. 2nd ed. Available from: http://apps.who.int/iris/bitstream/10665/144785/1/9789241548953_eng.pdf
59. Arrossi S, Temin S, Garland S, Eckert LON, Bhatla N, Castellsagué X, et al. Primary prevention of cervical cancer: American Society of Clinical Oncology resource-stratified guideline. J Glob Oncol. 2017:JGO. 2016.008151.
60. Horton S, Gauvreau CL. Cancer in low-and middle-income countries: an economic overview. In: Gelbrand H, Jha P, Sankaranarayanan R, et al., editors. Cancer: disease control priorities, vol. 3. Washington, DC: International Bank for Reconstruction and Development/World Bank; 2015. p. 263–80.
61. WHO. World Health Organization: new WHO guide to prevent and control cervical cancer. World Health Organization. 2014. Available from: http://www.who.int/mediacentre/news/releases/2014/preventing-cervical-cancer/en/
62. Fesenfeld M, Hutubessy R, Jit M. Cost-effectiveness of human papillomavirus vaccination in low and middle income countries: a systematic review. Vaccine. 2013;31(37):3786–804.
63. Castle PE, Jeronimo J, Temin S, Shastri SS. Screening to prevent invasive cervical cancer: ASCO resource-stratified clinical practice guideline. J Clin Oncol. 2017;35(11):1250–2.
64. IARC. IARC handbooks of cancer prevention, Cervix cancer screening, vol. 10. Lyon: International Agency for Cancer Research; 2005.
65. Bruni L, Diaz M, Barrionuevo-Rosas L, Herrero R, Bray F, Bosch FX, et al. Global estimates of human papillomavirus vaccination coverage by region and income level: a pooled analysis. Lancet Glob Health. 2016;4(7):e453–e63.
66. Markowitz LE, Dunne EF, Saraiya M, Chesson HW, Curtis CR, Gee J, et al. Human papillomavirus vaccination: recommendations of the advisory committee on immunization practices (ACIP). Methods. 2007;12:15.
67. Shastri SS, Mittra I, Mishra GA, Gupta S, Dikshit R, Singh S, et al. Effect of VIA screening by primary health workers: randomized controlled study in Mumbai, India. J Natl Cancer Inst. 2014;106(3):dju009.
68. Arbyn M, Verdoodt F, Snijders PJ, Verhoef VM, Suonio E, Dillner L, et al. Accuracy of human papillomavirus testing on self-collected versus clinician-collected samples: a meta-analysis. Lancet Oncol. 2014;15(2):172–83.
69. Arrossi S, Thouyaret L, Herrero R, Campanera A, Magdaleno A, Cuberli M, et al. Effect of self-collection of HPV DNA offered by community health workers at home visits on uptake of screening for cervical cancer (the EMA study): a population-based cluster-randomised trial. Lancet Glob Health. 2015;3(2):e85–94.
70. Castle PE, Rausa A, Walls T, Gravitt PE, Partridge EE, Olivo V, et al. Comparative community outreach to increase cervical cancer screening in the Mississippi Delta. Prev Med. 2011;52(6):452–5.

71. Adesina A, Chumba D, Nelson AM, Orem J, Roberts DJ, Wabinga H, et al. Improvement of pathology in sub-Saharan Africa. Lancet Oncol. 2013;14(4):e152–e7.
72. Sayed S, Lukande R, Fleming KA. Providing pathology support in low-income countries. Alexandria: American Society of Clinical Oncology; 2015.
73. Olmsted SS, Moore M, Meili RC, Duber HC, Wasserman J, Sama P, et al. Strengthening laboratory systems in resource-limited settings. Am J Clin Pathol. 2010;134(3):374–80.
74. Sankaranarayanan R, Anorlu R, Sangwa-Lugoma G, Denny LA. Infrastructure requirements for human papillomavirus vaccination and cervical cancer screening in sub-Saharan Africa. Vaccine. 2013;31:F47–52.
75. Chuang LT, Feldman S, Nakisige C, Temin S, Berek JS. Management and care of women with invasive cervical cancer: ASCO resource-stratified clinical practice guideline. J Clin Oncol. 2016;34(27):3354–5.
76. Jeronimo J, Castle PE, Temin S, Denny L, Gupta V, Kim JJ, et al. Secondary prevention of cervical cancer: ASCO resource-stratified clinical practice guideline. J Glob Oncol. 2016;3(5):635–57. JGO006577
77. HVO. Health voluunter overseas Washington, DC. 2017. Available from: http://www.hvousa.org
78. GHWA. Global Health Workforce Alliance. 2017. Available from: http://www.int/entity/workforcealliance/en
79. Chuang L, Moore KN, Creasman WT, Goodman A, Cooper HH, Price FV, et al. Teaching gynecologic oncology in low resource settings: a collaboration of health volunteers overseas and the society of gynecologic oncology: Elsevier; 2014.
80. IGCS. The gynecologic oncology global curriculum & mentorship program. Louisville: International Gynecologic Cancer Society; 2017. Available from: http://www.igcs.org
81. ASCO. International professional development opportunities. Alexandria: American Society of Clinical Oncology; 2017. Available from: http://www.asco.org
82. ASCO. Conquer cancer foundation supports oncologists and patients with cancer worldwide. ASCO Daily News, May 29, 2015:44 http://www.am.asco.org/dn
83. ASCO. Multidisciplinary cancer management courses improve cancer care worldwide. ASCO Daily News, May 31, 2015:35 http://www.am.asco.org/dn
84. Shulman LN, Trimble E, Enui A, editors. Recognizing global oncology as an academic field. In: Annual conference of the American Society of Clinical Oncology, Chicago, IL, USA. 2017.
85. Gao GF, Feng Y. On the ground in Sierra Leone. Science. 2014;346(6209):666.
86. Sachs JD. From millennium development goals to sustainable development goals. Lancet. 2012;379(9832):2206–11.
87. Mehl G, Labrique A. Prioritizing integrated mHealth strategies for universal health coverage. Science. 2014;345(6202):1284–7.
88. UNGAS. The UN general assembly, 67/81 Global Health and Foreign Policy. United Nations, New York. 2012.
89. WHO. World Health report: health systems financing: the path to universal coverage. Geneva. 2010.
90. Gupta V, Kerry VB, Goosby E, Yates R. Politics and universal health coverage – the post-2015 global health agenda. N Engl J Med. 2015;373(13):1189–92.
91. WHO. World Health Assembly Resolution WHA 58.22, May 2005. http://www.who.int/ipcs/publications/wha/cancer_resolution.pdf. Accessed 26 June 2017.
92. WHO. Global status report on non-communicable diseases 2010, Geneva. 2011.
93. WHO. Global Health Observatory data repository: NCD country capacity survey 2010. Geneva: World Health Organization; 2010. Available from: http://apps.who.int/gho/data/
94. Stefan DC, Elzawawy AM, Khaled HM, Ntaganda F, Asiimwe A, Addai BW, et al. Developing cancer control plans in Africa: examples from five countries. Lancet Oncol. 2013;14(4):e189–e95.
95. Reich MR, Shibuya K. The future of Japan's health system – sustaining good health with equity at low cost. N Engl J Med. 2015;373(19):1793–7.

96. WHO. Countdown to 2015 decade report. Nigeria: World Health Organization/World Bank; 2015. Available from: http://www.who.int
97. Reddy KS. India's aspirations for universal health coverage. N Engl J Med. 2015;373(1):1–5.
98. Sanou B. The world in 2013: ICT facts and figures. Geneva: International Telecommunications Union; 2013.
99. Organization WH. The partnership for maternal, newborn & child health. A global review of the key interventions related to reproductive, maternal, newborn and child health (Rmnch). Geneva: PMNCH; 2011. 2017
100. Labrique AB, Vasudevan L, Kochi E, Fabricant R, Mehl G. mHealth innovations as health system strengthening tools: 12 common applications and a visual framework. Glob Health Sci Pract. 2013;1(2):160–71.
101. Moobilevrs. Uganda Registration Services Bureau, Mobile Vital Records System. Available from: http://www.mobilevrs.co.ug/home.php
102. Campbell J, Buchan J, Cometto G, David B, Dussault G, Fogstad H, et al. Human resources for health and universal health coverage: fostering equity and effective coverage. Bull World Health Organ. 2013;91(11):853–63.
103. Mwanahamuntu MH, Sahasrabuddhe VV, Kapambwe S, Pfaendler KS, Chibwesha C, Mkumba G, et al. Advancing cervical cancer prevention initiatives in resource-constrained settings: insights from the cervical cancer prevention program in Zambia. PLoS Med 2011;8(5):e1001032.
104. Suri T, Jack W. The long-run poverty and gender impacts of mobile money. Science. 2016;354(6317):1288–92.
105. Corby N. Using mobile finance to reimburse sexual and reproductive health vouchers in Madagascar. London: Marie Stopes International; 2012.
106. Holmes D. Margaret Chan: committed to universal health coverage. Lancet. 2012;380(9845):879.
107. Fakunle B. Obio community health model "wipes away tears" in Nigeria [Internet]. VOICES: Ideas and Insight From Explorers, July 2, 2015.
108. Sudhinaraset M, Ingram M, Lofthouse HK, Montagu D. What is the role of informal healthcare providers in developing countries? A systematic review. PLoS One. 2013;8(2):e54978.
109. Das J, Holla A, Das V, Mohanan M, Tabak D, Chan B. In urban and rural India, a standardized patient study showed low levels of provider training and huge quality gaps. Health Aff. 2012;31(12):2774–84.
110. Powell-Jackson T. A pragmatic way forward? Science. 2016;354(6308):34–5.
111. GovernmentofIndia. Ministry of Health and Family Welfare, Government of India, National Health Policy 2015 Draft, December 2014. Available from: http://www.nhp.gov.in/sites/default/files/pdf/draft_national_health_policy_2015.pdf
112. Mackintosh M, Channon A, Karan A, Selvaraj S, Cavagnero E, Zhao H. What is the private sector? Understanding private provision in the health systems of low-income and middle-income countries. Lancet. 2016;388(10044):596–605.
113. Das J, Chowdhury A, Hussam R, Banerjee AV. The impact of training informal health care providers in India: a randomized controlled trial. Science. 2016;354(6308):aaf7384.
114. Haines A, Kuruvilla S, Borchert M. Bridging the implementation gap between knowledge and action for health. Bull World Health Organ. 2004;82(10):724–31.
115. Cockcroft A, Masisi M, Thabane L, Andersson N. Legislators learning to interpret evidence for policy. Science. 2014;345(6202):1244–5.
116. Evipnet. Report on International Forum on evidence informed health policy in low- and middle-income countries, Addis Ababa, Ethiopia, 27 to 31 August 2012. http://global.evipnet.org/wp-content/uploads/2013/02/Addisreport2012.pdf
117. Tovey CA. In defense of basic research. American Association for the Advancement of Science. 2017.
118. Flexner A. The usefulness of useless knowledge. Princeton: Princeton University Press; 2017.

119. Alberts B. Editorial. Science for life. Science. 2017;355(6332):1353.
120. Mgwebi T. Three Q's. 2015. Available from: http://scim.ag/MgwebiAAAS
121. Nattrass N. Mortal combat: AIDS denialism and the struggle for antiretrovirals in South Africa. Scottsville: University of KwaZulu-Natal Press; 2007.
122. Jamison DT, Breman JG, Measham AR, Alleyne G, Claeson M, Evans DB, et al. Disease control priorities in developing countries. New York/Washington, DC: Oxford University Press/World Bank; 2006.
123. Carney JP. Science advocacy, defined. Science. 2014;345(6194):243.
124. Houston OR. We have a narrative: why science needs story. Chicago: University of Chicago Press; 2015.
125. Luna RE. The storytelling scientist. Washington, DC: American Association for the Advancement of Science; 2015. p. 391.
126. NAN. Schools nationwide to teach maths, science subjects in indigenous languages – minister. Punch, May 31, 2017.
127. Akhampapa. Nigerian schools will start teaching maths, science in local dialect soon – minister 2017. Available from: http://www.naijaloaded.com.ng/author/akhampapa/
128. Scagliotti GV, Parikh P, Von Pawel J, Biesma B, Vansteenkiste J, Manegold C, et al. Phase III study comparing cisplatin plus gemcitabine with cisplatin plus pemetrexed in chemotherapy-naive patients with advanced-stage non–small-cell lung cancer. J Clin Oncol. 2008;26(21):3543–51.
129. Lee BG. KIST at 50, beyond the miracle. American Association for the Advancement of Science. 2016.
130. Mtingwa SK, Winick H. SESAME and beyond. Washington, DC American Association for the Advancement of Science. 2017.
131. Pastrana SJ. Science in US-Cuba relations. Science. 2015;348(6236):735.
132. Finlay CJ. Yellow fever: Oliver and Boyd; 1894.
133. Stone R. In from the cold. After keeping science alive during decades of scarcity, Cuba's "guerilla scientists" are ready to rejoin the world. Science. 2015;348(6236):746–51.
134. Singapore. Singapore boosts science. Science. 2016;351(6270):209.
135. NewDelhi. India eyes Mars and Venus. Science. 2017;355(6327):779. http://scim.ag/Indiaastro
136. Niemeier D, Gombachika H, Richards-Kortum R. How to transform the practice of engineering to meet global health needs. Science. 2014;345(6202):1287–90.
137. Horton R, Beaglehole R, Bonita R, Raeburn J, McKee M, Wall S. From public to planetary health: a manifesto. Lancet. 2014;383(9920):847.
138. IOM. Institute of Medicine. Evaluation of *PEPFAR*. Washington, DC: National Academies Press; 2013.
139. Baldinger P, Ratterman W. Powering health: options for improving energy services at health facilities in Ethiopia. Washington, DC: USAID; 2008.
140. Frost LJ, Reich MR. Access: how do good health technologies get to poor people in poor countries?: Harvard Center for Population and Development Studies; 2008.
141. Richards-Kortum R, Gray LV, Oden M. Engaging undergraduates in global health technology innovation. Science. 2012;336(6080):430–1.
142. Aker JC, Mbiti IM. Mobile phones and economic development in Africa. J Econ Perspect. 2010;24(3):207–32.
143. Sciencemag. Low-cost x-rays for all. Science. 2015;347(6227):1180–1.
144. Stone R. Graying Cuba strains socialist safety net. Science. 2015;348(6236):750.
145. Holmes J, Jones B, Heap B. Smart villages. Science. 2015;350(6259):359.
146. Sambo AS. Strategic developments in renewable energy in Nigeria. International Association for Energy Economics. 2009;16(3):15–9.
147. Wu Z. HIV/AIDS in China: beyond the numbers: Springer; 2017.
148. Harper K. The aftermath of AIDS in China: American Association for the Advancement of Science; 2016.

149. Sporn MB. The war on cancer. Lancet. 1996;347(9012):1377–81.
150. Kesselheim AS, Avorn J. New "21st century cures" legislation: speed and ease vs science. JAMA. 2017;317(6):581–2.
151. Kaiser J, Couzin-Frankel J. Biden seeks clear course for his cancer moonshot. Science. 2016;351(6271):325–6.
152. Singer DS, Jacks T, Jaffee E. A US "Cancer moonshot" to accelerate cancer research. Science. 2016;353(6304):1105–6.
153. Cavallo J. Cancer moonshot blue ribbon panel recommends 10 ways to speed cancer advances. The ASCO Post. 2016 September 25;2016:72.
154. Corsi DJ, Boyle MH, Lear SA, Chow CK, Teo KK, Subramanian S. Trends in smoking in Canada from 1950 to 2011: progression of the tobacco epidemic according to socioeconomic status and geography. Cancer Causes Control. 2014;25(1):45–57.
155. StatisticsCanada. Statistics Canada 2011. Canadian Community Health Survey (CCHS), Cycle 1.1, 2.1, 3.1, 4.1 and 2008, 2009, 2010 Annual Component surveys (Microdata. Ottawa, Ontario: Statistics Canada Health Statistics Division (producer and distributor): Ottawa Ontario: Data Liberation Initiative (distributor); 2011.
156. CancerCareOntario. Cancer Fact: Lung Cancer accounts for largest proportion of premature Ontario cancer deaths. Available from: http://www.cancercare.on.ca/cancerfacts.
157. Field JK, Oudkerk M, Pedersen JH, Duffy SW. Prospects for population screening and diagnosis of lung cancer. Lancet. 2013;382(9893):732–41.
158. Sox HC. Better evidence about screening for lung cancer. Mass Medical Soc. 2011.
159. Moyer VA. Screening for lung cancer: US preventive services task force recommendation statement. Ann Intern Med. 2014;160(5):330–8.
160. McMahon PM, Kong CY, Bouzan C, Weinstein MC, Cipriano LE, Tramontano AC, et al. Cost-effectiveness of computed tomography screening for lung cancer in the United States. J Thorac Oncol. 2011;6(11):1841–8.
161. Pyenson BS, Sander MS, Jiang Y, Kahn H, Mulshine JL. An actuarial analysis shows that offering lung cancer screening as an insurance benefit would save lives at relatively low cost. Health Aff. 2012;31(4):770–9.
162. Mahadevia PJ, Fleisher LA, Frick KD. Eng J, Goodman SN, Powe NR. Lung cancer screening with helical computed tomography in older adult smokers: a decision and cost-effectiveness analysis. JAMA. 2003;289(3):313–22.
163. Goffin JR, Flanagan WM, Miller AB, Fitzgerald NR, Memon S, Wolfson MC, et al. Cost-effectiveness of lung cancer screening in Canada. JAMA Oncol. 2015;1(6):807–13.
164. Raymakers AJ, Mayo J, Lam S, Fitzgerald JM, Whitehurst DG, Lynd LD. Cost-effectiveness analyses of lung cancer screening strategies using low-dose computed tomography: a systematic review. Appl Health Econ Health Policy. 2016;14(4):409–18.
165. Fauci AS, Marston HD. Ending the HIV–AIDS pandemic – follow the science. N Engl J Med. 2015;373(23):2197–9.
166. Molina J-M, Capitant C, Spire B, Pialoux G, Cotte L, Charreau I, et al. On-demand preexposure prophylaxis in men at high risk for HIV-1 infection. N Engl J Med. 2015;373(23):2237–46.
167. START-GROUP. Initiation of antiretroviral therapy in early asymptomatic HIV infection. N Engl J Med. 2015;2015(373):795–807.
168. Grinsztejn B, Hosseinipour MC, Ribaudo HJ, Swindells S, Eron J, Chen YQ, et al. Effects of early versus delayed initiation of antiretroviral treatment on clinical outcomes of HIV-1 infection: results from the phase 3 HPTN 052 randomised controlled trial. Lancet Infect Dis. 2014;14(4):281–90.
169. El-Sadr WM, Harripersaud K, Bayer R. End of AIDS – hype versus hope. Science. 2014;345(6193):166.
170. Barouch DH, Deeks SG. Immunologic strategies for HIV-1 remission and eradication. Science. 2014;345(6193):169–74.
171. Finzi D, Hermankova M, Pierson T, Carruth LM, Buck C, Chaisson RE, et al. Identification of a reservoir for HIV-1 in patients on highly active antiretroviral therapy. Science. 1997;278(5341):1295–300.

172. Persaud D, Zhou Y, Siliciano JM, Siliciano RF. Latency in human immunodeficiency virus type 1 infection: no easy answers. J Virol. 2003;77(3):1659–65.
173. Chun T-W, Stuyver L, Mizell SB, Ehler LA, Mican JAM, Baseler M, et al. Presence of an inducible HIV-1 latent reservoir during highly active antiretroviral therapy. Proc Natl Acad Sci. 1997;94(24):13193–7.
174. Chun T-W, Engel D, Berrey MM, Shea T, Corey L, Fauci AS. Early establishment of a pool of latently infected, resting CD4+ T cells during primary HIV-1 infection. Proc Natl Acad Sci. 1998;95(15):8869–73.
175. Chun T-W, Carruth L, Finzi D, Shen X. Quantification of latent tissue reservoirs and total body viral load in HIV-1 infection. Nature. 1997;387(6629):183–8.
176. Chun T-W, Davey RT, Engel D, Lane HC, Fauci AS. AIDS: re-emergence of HIV after stopping therapy. Nature. 1999;401(6756):874–5.
177. Finzi D, Blankson J, Siliciano JD, Margolick JB, Chadwick K, Pierson T, et al. Latent infection of CD4+ T cells provides a mechanism for lifelong persistence of HIV-1, even in patients on effective combination therapy. Nat Med. 1999;5(5):512–7.
178. Ho Y-C, Shan L, Hosmane NN, Wang J, Laskey SB, Rosenbloom DI, et al. Replication-competent noninduced proviruses in the latent reservoir increase barrier to HIV-1 cure. Cell. 2013;155(3):540–51.

Index

C. K. O. Williams, *Cancer and AIDS*,
https://doi.org/10.1007/978-3-319-99238-9